Essential Mathematics for Life

BOOK 5

Geometry

Fourth Edition

GLENCOE
McGraw-Hill

New York, New York
Columbus, Ohio
Woodland Hills, California
Peoria, Illinois

Photo credits: Cover, © Ralph Mercer/Tony Stone Images; 55, Michele Wigginton; 75, Cobalt; 92, Glencoe file; 117, Will Faller; 130, Alain Choisnet/Image Bank; 136, Larry Hamill; 199, Glencoe file; 211, Eric Schweikardt/Image Bank

Send all inquiries to:
Glencoe/McGraw-Hill
936 Eastwind Drive
Westerville, Ohio 43081

ISBN: 0-02-802613-6

4 5 6 7 8 9   066   02 01 00 99

Authors

Mary S. Charuhas
Associate Dean
College of Lake County
Grayslake, Illinois

Dorothy McMurtry
District Director of ABE,GED,
  ESL
City Colleges of Chicago
Chicago, Illinois

The Mathematics Faculty
American Preparatory Institute
Killeen, Texas

Contributing Writers

Kathryn S. Harr
Mathematics Instructor
Pickerington, Ohio

Priscilla Ware
Educational Consultant and
  Instructor
Columbus, Ohio

Dr. Pearl Chase
Professional Consultants of Dallas
Cedar Hill, Texas

Contributing Editors and Reviewers

Barbara Warner
Monroe Community College
Rochester, New York

Anita Armfield
York Technical College
Rock Hill, South Carolina

Judy D. Cole
Lafayette Regional Technical
  Institute
Lafayette, Louisiana

Mary Fincher
New Orleans Job Corps
New Orleans, Louisiana

Cheryl Gunderson
Rusk Community Learning
  Center
Ladysmith, Wisconsin

Cynthia A. Love
Columbus City Schools
Columbus, Ohio

Joyce Claar
South Westchester BOCES
Valhalla, New York

John Grabowski
St. Joseph Hill Academy
Staten Island, New York

Virginia Victor
Maple Run Youth Center
Cumberland, Maryland

Sandi Braga
College of South Idaho
Twin Falls, Idaho

Maggie Cunningham
Adult Education
Schertz, Texas

Sylvia Gilliard
Naval Consolidated Brig
Charleston, South Carolina

Eva Eaton-Smith
Cecil Community College
Elkton, Maryland

Fabienne West
John C. Calhoun State
  Community College
Decatur, Alabama

C O N T E N T S

Geometry

Unit 1 Review of Basic Arithmetic

Unit 2 Angles

Unit 3 Triangles

Unit 4 Polygons

Unit 5 Quadrilaterals

Unit 6 Circles

Unit 7 Coordinate Geometry

Unit 8 Solid Geometry

Unit 9 Cylinders and Cones

Unit 10 Prisms and Pyramids

Unit 11 Spheres

1

Review of Basic Arithmetic

Problem Solving

Read each problem and circle the correct answer.

1. A board $3\frac{3}{4}$ feet long was cut from a 10-foot board. How many feet are left on the board?

   (1) $7\frac{3}{4}$ feet   **(2)** $6\frac{1}{4}$ feet

   (3) $13\frac{3}{4}$ feet   **(4)** $6\frac{3}{4}$ feet

   (5) $7\frac{1}{4}$ feet

2. Derrick used 3 gallons of paint to cover 1,350 square feet. How many gallons would he need to paint 1,800 square feet?

   (1) 1.33 gallons   **(2)** 450 gallons

   (3) 2.25 gallons   **(4)** 4 gallons

   (5) More information is needed.

3. If 75 cm of wire is cut from a roll of wire 50 m long, how much wire is left on the roll?

   (1) 49.25 m   **(2)** 25 m

   (3) 49.75 m   **(4)** 1.5 m

   (5) 50.75 m

4. The top of a table is $1\frac{3}{16}$ inches thick. The legs of the table are $27\frac{5}{8}$ inches long. How tall is the table?

   (1) $28\frac{8}{24}$ inches   **(2)** $26\frac{2}{8}$ inches

   (3) $28\frac{13}{16}$ inches   **(4)** $28\frac{1}{2}$ inches

   (5) $26\frac{7}{16}$ inches

5. Yao has 100 feet of jute. How many macrame flowerpot hangers can he make if each hanger takes $12\frac{1}{2}$ feet of jute?

   (1) 8 hangers   **(2)** 1,250 hangers

   (3) 8.3 hangers   **(4)** 58.3 hangers

   (5) $\frac{1}{8}$ hangers

6. Use $C = \frac{5}{9}(F - 32)$ to change 72°F to a Celsius reading.

   (1) 40°C   **(2)** 12°C

   (3) $22\frac{2}{9}$°C   **(4)** 72°C

   (5) 161.6°C

7. The formula for finding board feet is Board feet = LWT where L is length in feet, W is width in feet, and T is thickness in inches. Find the number of board feet in a piece of lumber that measures 12 feet long, 10 inches wide, and 2 inches thick.

   (1) 240 board feet

   (2) 24 board feet

   (3) $1\frac{2}{3}$ board feet

   (4) 2,880 board feet

   (5) 20 board feet

8. Petra and Alonzo are making a table with a top that measures $3\frac{3}{4}$ feet long by $2\frac{1}{2}$ feet wide. What is the perimeter of the tabletop? Use $P = 2l + 2w$.

   (1) $5\frac{2}{3}$ feet

   (2) $12\frac{1}{2}$ feet

   (3) $6\frac{1}{4}$ feet

   (4) $10\frac{2}{3}$ feet

   (5) $35\frac{1}{2}$ feet

9. Find the area of the top of the table described in Problem 8. Use $A = lw$.

   (1) $6\frac{3}{8}$ ft²   **(2)** $18\frac{3}{8}$ ft²

   (3) $1\frac{1}{2}$ ft²   **(4)** $9\frac{3}{8}$ ft²

   (5) $\frac{2}{3}$ ft²

10. The area of a circle is given by the formula $A = \pi r^2$. Find the area of a circle with a radius of 3 inches. Use $\pi = 3.14$.

    (1) 9.42 in.²   **(2)** 18.84 in.²

    (3) 28.26 in.²   **(4)** 88.7364 in.²

    (5) 3.0692 in.²

Operations With Whole Numbers

The operations—addition, subtraction, multiplication, and division—are often applied to solve problems in geometry.

Addition:
Addition is indicated by certain clue words. Some clue words you may find are **total, sum, increase, together, both,** and **altogether.**

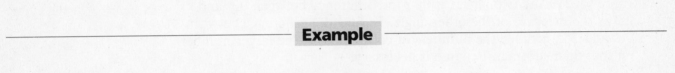

Example

A. Find the perimeter of the figure shown in the diagram.

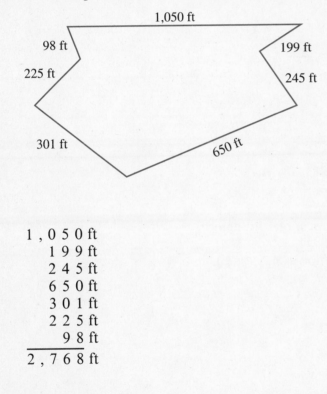

```
1 , 0 5 0 ft
    1 9 9 ft
    2 4 5 ft
    6 5 0 ft
    3 0 1 ft
    2 2 5 ft
      9 8 ft
─────────────
2 , 7 6 8 ft
```

The perimeter of a figure is the sum of its sides. To find the perimeter of this figure, arrange the numbers in columns and add. Start at the right with the ones column.

Check: Add in the opposite direction. That is, if you added from the top to the bottom, check by adding from the bottom to the top.

Subtraction:
Subtraction is indicated by these clue words: **difference, remainder, how much more, how much less, what is left, decreased by, diminished by,** and **less than.**

Example

B. **Complementary angles** are two angles whose sum is 90°. If one complementary angle measures 25°, what is the measure of the other angle?

If the sum of the two angles is 90°, the difference between 90° and 25° is the measure of the other angle. To find the difference between 90° and 25°, arrange the numbers in columns with the larger number on top, then subtract. Rename if necessary.

$$
\begin{array}{r}
{}^{8}\!\!\!\not{9}\,{}^{1}0° \\
-\ 2\ 5° \\
\hline
5°
\end{array}
\qquad
\begin{array}{r}
{}^{8}\!\!\!\not{9}\,{}^{1}0° \\
-\ 2\ 5° \\
\hline
6\ 5°
\end{array}
\qquad
\text{Check:}
\qquad
\begin{array}{r}
{}^{1}6\ 5° \\
+\ 2\ 5° \\
\hline
9\ 0°
\end{array}
$$

Multiplication:
Some clue words for multiplication are **product, times, of,** and **apiece.**

Multiplication is indicated in equations by the symbols **×**, **·**, or **( )**.

6×4

$6 \cdot 4$

$6(4)$

$(6)(4)$

MATH HINT

To multiply a number by 10, 100, 1000, etc., place zeros after the number. For example, $34 \cdot 10 = 340$, $34 \cdot 100 = 3400$, and $34 \cdot 1000 = 34,000$.

4

C. The area of a rectangle is found by multiplying its length times its width. The formula is $A = lw$ where A is the area, l is the length, and w is the width.

Find the area of a rectangular room that measures 12 feet by 9 feet.

$A = lw$ or $A = 12(9)$

To find the product of 12 and 9, write the larger number above the smaller number and multiply.

$$
\begin{array}{r}
1\,2 \text{ ft} \\
\times \quad 9 \text{ ft} \\
\hline
1\,0\,8 \text{ sq ft}
\end{array}
$$

Division:
Some clue words for division are: **quotient, how many times, average, shared equally, cut,** and **split.**

D. The **diameter** of a circle is two times the radius. A circle has a diameter of 50 in. Find the radius.

Use the formula, $r = \frac{d}{2}$, where r is the radius and d is the diameter. Write the number being divided under the division sign.

$$
2\overline{)50} \qquad
\begin{array}{r}
2 \\
2\overline{)50} \\
4 \\
\hline
10
\end{array}
\qquad
\begin{array}{r}
25 \\
2\overline{)50} \\
4 \\
\hline
10 \\
10 \\
\hline
\end{array}
\qquad \text{Check:} \qquad
\begin{array}{r}
2\,5 \text{ in.} \\
\times \quad 2 \\
\hline
5\,0 \text{ in.}
\end{array}
$$

The radius is 25 in.

The sum of the three angles of any triangle is 180°. Find the sum of each set of numbers and tell whether the numbers can be the angles of a triangle.

1. 45°, 45°, 89° _____

2. 60°, 40°, 70° _____

3. 35°, 55°, 90°   _____   **4.** 38°, 66°, 76°   _____

_____   _____

5. 15°, 17°, 158°   _____   **6.** 29°, 35°, 85°   _____

_____   _____

Problem Solving

Solve the following problems.

7. Two angles of a triangle measure 80° and 15°. Find the measure of the third angle.

_____

8. A pasture measures 625 feet by 450 feet. Find the area of the pasture.

_____

9. To change inches to feet, divide by 12. Change 1,346 inches to feet.

_____

10. To change feet to inches, multiply by 12. Change 6 feet to inches.

_____

11. A pasture measures 625 feet on each of the two longer sides and 450 feet on each of the two shorter sides. What is the perimeter of the pasture?

_____

Operations With Fractions

A **fraction** is a part of a whole.

$\frac{1}{2}$ ← numerator
← denominator

The **numerator** (top number) tells the number of parts used. The **denominator** (bottom number) tells how many parts the whole has been divided into.

Types of fractions:

Proper fraction. The numerator is smaller than the denominator. It is less than the whole.

$\frac{1}{2}$ $\frac{2}{3}$ $\frac{9}{47}$

Improper fraction. The numerator is larger than or equal to the denominator. It is equal to or greater than the whole.

$\frac{3}{2}$ $\frac{2}{1}$ $\frac{7}{7}$

Mixed number. The sum of a whole number and a fraction.

$3\frac{3}{4}$ $1\frac{1}{8}$ $97\frac{2}{3}$

To compare fractions, the fractions must be written with common denominators. One way to find a common denominator is to take the largest denominator and list its multiples. When you are able to divide a multiple by the other denominators evenly, you have found a common denominator.

Examples

A. Find a common denominator for these fractions: $\frac{3}{8}, \frac{5}{12}, \frac{7}{16}$

The largest denominator is 16.
List its multiples and check.

$16 \times 1 = 16$
  16 can be divided evenly by **8** but not by **12**.
$16 \times 2 = 32$
  32 can be divided evenly by **8** but not by **12**.
$16 \times 3 = 48$
  48 can be divided evenly by **8** and **12** as well as by 16.

The common denominator is 48.

To change a fraction into one having the common denominator, multiply both its numerator and denominator by the number needed to change the denominator to the common denominator.

$$\frac{3}{8} = \frac{3 \times 6}{8 \times 6} = \frac{18}{48}$$
$$\frac{5}{12} = \frac{5 \times 4}{12 \times 4} = \frac{20}{48}$$
$$\frac{7}{16} = \frac{7 \times 3}{16 \times 3} = \frac{21}{48}$$

B. If a carpenter needs nails longer than $1\frac{3}{8}$ inches, should the carpenter use $1\frac{3}{4}$-inch or $1\frac{5}{16}$-inch nails?

Change the fractions to equivalent fractions with a common denominator. The lowest common denominator is 16.

$$1\frac{3}{8} = 1\frac{6}{16} \qquad\qquad 1\frac{3}{4} = 1\frac{12}{16} \qquad\qquad 1\frac{5}{16} = 1\frac{5}{16}$$
$$1\frac{6}{16} < 1\frac{12}{16} \qquad\qquad 1\frac{6}{16} > 1\frac{5}{16}$$

Since $1\frac{3}{4} > 1\frac{3}{8}$, the carpenter needs to use the $1\frac{3}{4}$-inch nails.

To change an improper fraction to a whole or mixed number, divide the denominator into the numerator. Write any remainder over the original denominator and reduce if possible.

$$\frac{1\frac{6}{9} = 1\frac{2}{3}}{\frac{15}{9} \rightarrow 9\overline{)15}}$$
$$\frac{9}{6}$$

To reduce a proper fraction, divide the numerator and the denominator by the largest number that evenly divides into both.

$$\frac{18 \div 6}{24 \div 6} = \frac{3}{4}$$

If the largest number was not used, keep dividing until no number greater than 1 will divide evenly into both the numerator and the denominator.

$$\frac{18 \div 2}{24 \div 2} = \frac{9 \div 3}{12 \div 3} = \frac{3}{4}$$

To add fractions, follow these steps:

Step 1   Change to fractions with a common denominator.

Step 2   Add the numerators and write the sum over the common denominator.

Step 3   If the fraction in the answer is improper, write it as a mixed or whole number and combine with the whole number.

Step 4   Reduce the remaining fraction if possible.

C. Find the perimeter of a triangle with sides measuring $2\frac{5}{8}$ inches, $3\frac{3}{4}$ inches, and $5\frac{1}{4}$ inches.

Find the common denominator. Add the whole numbers and the fractions.

$$2\frac{5}{8} = 2\frac{5}{8}$$
$$3\frac{3}{4} = 3\frac{6}{8}$$
$$+\;5\frac{1}{4} = 5\frac{2}{8}$$

$$1\,0\,\tfrac{13}{8} = 10 + 1\frac{5}{8} \qquad \text{Change the improper fraction to a}$$
$$= 11\frac{5}{8} \qquad\qquad \text{mixed number.}$$

To subtract fractions, follow these steps:

Step 1 Change to fractions with a common denominator.

Step 2 Rename (borrow) if necessary.

Step 3 Subtract the numerators and write the difference over the common denominator.

Step 4 Reduce the answer if possible.

D. A machinist must be able to interpret drawings involving dimensions written as fractions. Fractions often must be added to find total length or subtracted to find missing dimensions.

Find the missing dimension of this bolt.

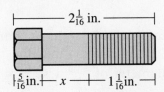

Find the sum of the two known dimensions.

$$\frac{5}{16}$$
$$+\;1\frac{1}{16}$$
$$1\frac{6}{16}$$

Subtract the sum of the known dimensions from the total length to find the missing dimension.

$$2\frac{1}{16} = 1\frac{17}{16}$$
$$-\;1\frac{6}{16} = 1\frac{6}{16}$$
$$\frac{11}{16}$$

The missing dimension of this bolt is $\frac{11}{16}$ inches.

To change a whole number to an improper fraction, choose a number (other than 0) for the denominator. Then find the numerator by multiplying the whole number by the number chosen as the denominator.

$$6 = \frac{6 \times 1}{1} = \frac{6}{1} \qquad 1 = \frac{1 \times 12}{12} = \frac{12}{12} \qquad 3 = \frac{3 \times 16}{16} = \frac{48}{16}$$

To change a mixed number to an improper fraction, multiply the whole number by the denominator of the fraction and then add the numerator. Write this total over the original denominator.

--- **Examples** ---

E. Change $3\frac{1}{7}$ to an improper fraction.

$$3\frac{1}{7} = \frac{(3 \times 7) + 1}{7} = \frac{22}{7}$$

F. Change 4 to an improper fraction with 12 as the denominator.

$$4 = \frac{4 \times 12}{12} = \frac{48}{12}$$

To multiply fractions or mixed numbers, follow these steps:

Step 1   Change any mixed or whole numbers to improper fractions.

Step 2   Multiply numerators by numerators and denominators by denominators.

Step 3   Reduce the answer if possible.

--- **Examples** ---

G. The circumference, or the distance around a circle, is found by multiplying π times the diameter of the circle. Find the circumference of a circle with a diameter of $5\frac{1}{4}''$. Use $\pi = \frac{22}{7}$.

$$C = \pi d$$
$$C = \frac{22}{7} \cdot 5\frac{1}{4}$$
$$C = \frac{22}{7} \cdot \frac{21}{4}$$
$$C = \frac{462}{28}$$
$$C = 16\frac{1}{2}''$$

Problems often can be made easier by **cancellation.** This means dividing a numerator and a denominator in the problem by the same number.

H. Use cancellation to solve $2\frac{1}{4} \times 2\frac{2}{3}$.

$$2\frac{1}{4} \cdot 2\frac{2}{3} = \frac{\overset{3}{\cancel{9}}}{\underset{1}{\cancel{4}}} \cdot \frac{\overset{2}{\cancel{8}}}{\underset{1}{\cancel{3}}}$$

$$= \frac{3 \cdot 2}{1 \cdot 1}$$

$$= \frac{6}{1}$$

$$= 6$$

To find the reciprocal of a fraction, exchange the positions of the numerator and denominator. This is called **inverting.**

The reciprocal of $\frac{3}{4}$ is $\frac{4}{3}$.

Since $3\frac{3}{4} = \frac{15}{4}$, the reciprocal of $3\frac{3}{4}$ is $\frac{4}{15}$.

Since $9 = \frac{9}{1}$, the reciprocal of 9 is $\frac{1}{9}$.

To divide fractions or mixed numbers, follow these steps:

Step 1   Change any mixed or whole number to an improper fraction.

Step 2   Invert the divisor (the number **after** the division sign).

Step 3   **Multiply** the fractions. Cancel if possible.

Step 4   Reduce the answer if possible.

───────────────── **Example** ─────────────────

I. If the circumference of a circle is known, the diameter can be found by dividing the circumference by π. Find the diameter of a circle with a circumference of 42 feet. Use $\pi = 3\frac{1}{7}$.

$$d = C \div \pi$$

$$d = 42 \div 3\frac{1}{7}$$

$$d = \frac{42}{1} \div \frac{22}{7}$$

$$d = \frac{\overset{21}{\cancel{42}}}{1} \times \frac{7}{\underset{11}{\cancel{22}}}$$

$$d = \frac{147}{11}$$

$$d = 13\frac{4}{11} \text{ feet}$$

Refer to the rectangle below to solve problems 1–3.

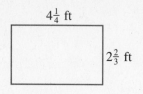

4¼ ft

2⅔ ft

1. The opposite sides of a rectangle are equal. This means that the rectangle has two sides measuring $4\frac{1}{4}$ feet. Find the total length of these two sides.

_____

2. Find the total length of the two sides measuring $2\frac{2}{3}$ feet.

_____

3. Find the perimeter of the rectangle.

_____

Problem Solving

Solve problems 4–10.

4. The area of a triangle is given by this formula: $A = \frac{1}{2}bh$ where A is the area, b is the base, and h is the height. Find the area of a triangle having a base of $5\frac{1}{2}$ inches and a height of $2\frac{2}{3}$ inches.

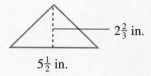

$2\frac{2}{3}$ in.

$5\frac{1}{2}$ in.

_____

5. Find the missing dimension.

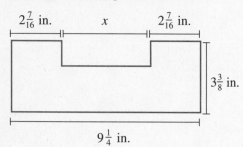

$2\frac{7}{16}$ in. x $2\frac{7}{16}$ in.

$3\frac{3}{8}$ in.

$9\frac{1}{4}$ in.

6. Find the missing dimension.

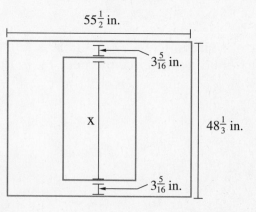

$55\frac{1}{2}$ in.

$3\frac{5}{16}$ in.

X

$48\frac{1}{3}$ in.

$3\frac{5}{16}$ in.

_____

_____

7. The taper of a piece of work is the difference between the diameter of the large end and the diameter of the small end. Find the taper of this piece.

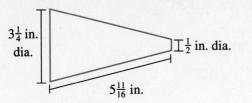

$3\frac{1}{4}$ in. dia.

$\frac{1}{2}$ in. dia.

$5\frac{11}{16}$ in.

_____

8. Find the taper of this piece of work.

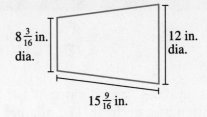

$8\frac{3}{16}$ in. dia.

12 in. dia.

$15\frac{9}{16}$ in.

_____

9. About how many floorboards $2\frac{1}{4}$ inches wide will be needed for a hallway $42\frac{1}{2}$ inches wide?

_____

10. One cubic foot of water weighs $62\frac{1}{2}$ pounds. What is the weight of $2\frac{1}{2}$ cubic feet of water?

_____

Operations With Decimals

When a fraction has a denominator of 10; 100; 1,000; 10,000; and so on, it can be written as a **decimal.** The number of decimal places indicates what power of ten is in the denominator.

This chart shows place values. Notice, there is no units (ones) place to the right of the decimal point.

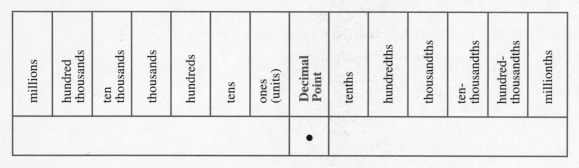

millions	hundred thousands	ten thousands	thousands	hundreds	tens	ones (units)	Decimal Point	tenths	hundredths	thousandths	ten-thousandths	hundred-thousandths	millionths
							•						

Mixed numbers are written with the whole number to the left of the decimal point and the fraction to the right. The decimal point is read as "and."

3.14 is read "three and fourteen hundredths."

If a number does not contain a decimal point, one can be written at the end of the number to the right.

25 = 25. = 25.0

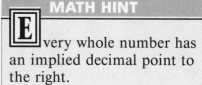

MATH HINT

Every whole number has an implied decimal point to the right.

To round decimals, follow these steps:

Step 1   Underline the number in the place to be rounded to.

Step 2   If the number to the right is 5 or more, add 1 to the underlined number.

Step 3   If the number to the right is less than 5, leave the underlined number as is.

Step 4   Drop the numbers to the right of the underlined number.

A. The number for π is a decimal number that never ends. π written to nine places is 3.141592654. For most problems, π can be rounded to two places, or hundredths. Round π to hundredths.

3.14̲1592654   Underline the 4 in the hundredths place. Since 1 is less than 5, the decimal rounds to 3.14.

To add or subtract decimals, follow these steps:

Step 1   Line up the decimal points underneath each other.

Step 2   Add or subtract decimals as you do whole numbers.

Step 3   Bring the decimal point straight down into the answer.

B. Find the perimeter of this figure.

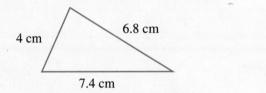

4 cm

6.8 cm

7.4 cm

$$
\begin{array}{r}
6\,.\,8 \\
4\,.\,0 \\
7\,.\,4 \\
\hline
1\,8\,.\,2
\end{array}
$$

The perimeter is 18.2 cm.

> **MATH HINT**
>
> **Z**eros can be added to the right of the last digit after the decimal point without changing the value of the decimal.

C. Find the missing dimension.

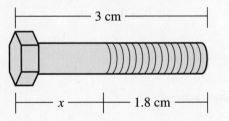

3 cm

x 1.8 cm

$$
\begin{array}{r}
3 \\
-\,1\,.\,8
\end{array}
\qquad
\begin{array}{r}
\overset{2}{\cancel{3}}\,.\,\overset{1}{0} \\
-\,1\,.\,8 \\
\hline
1\,.\,2
\end{array}
$$

The missing dimension is 1.2 cm.

To multiply decimals, follow these steps:

Step 1   Multiply decimals as you do whole numbers.

Step 2   Count the number of places to the right of the decimal points in the numbers you are multiplying.

Step 3   Put the total number of places from step 2 in your answer. Count from right to left. Add zeros on the left if necessary.

D. The circumference, or the distance around a circle, is found by multiplying π times the diameter of the circle. The formula is $C = \pi d$. Find the circumference of a circle with a diameter of 5.04 meters. Use $\pi = 3.14$.

$$
\begin{array}{r}
3.14 \\
\times\, 5.04 \\
\hline
1256 \\
15700 \\
\hline
15.8256
\end{array}
$$

There are 2 places to the right of the decimal.
There are 2 places to the right of the decimal.

Count 4 places starting from the "6" and moving right to left. Place the decimal point.

The circumference is 15.8256 meters.

To divide a decimal by a whole number, follow these steps:

Step 1   Put the answer decimal point straight above the decimal point in the dividend.

Step 2   Divide as with whole numbers.

$$
\begin{array}{r}
6.68 \\
4\overline{)26.72} \\
24 \\
\hline
27 \\
24 \\
\hline
32 \\
32 \\
\hline
\end{array}
\qquad
\begin{array}{r}
.03 \\
125\overline{)3.75} \\
3\,75 \\
\hline
\end{array}
$$

Remember, a decimal point can be put to the right of a whole number. Zeros can be added after the decimal point.

To divide a decimal by a decimal, follow these steps:

Step 1   Make the divisor a whole number by moving the decimal point to the right of the last digit.

Step 2   Move the decimal point in the dividend to the right the same number of places.

Step 3   Put the answer decimal point straight above the decimal in the dividend.

$$
\begin{array}{r}
2.46 \\
1.25\overline{)3.07.50} \\
2\,50 \\
\hline
57\,5 \\
50\,0 \\
\hline
7\,50 \\
7\,50 \\
\hline
\end{array}
\qquad
\begin{array}{r}
3\,750. \\
.004\overline{)15.000.} \\
12 \\
\hline
3\,0 \\
2\,8 \\
\hline
20 \\
20 \\
\hline
\end{array}
$$

MATH HINT

In the answer, a zero must be put in the ones place as a placeholder.

If an answer in a division problem is to be rounded to a certain place, carry the division out one place past the one to be rounded to.

Example

E. If the circumference of a circle is known, the diameter can be found by dividing the circumference by π. Find the diameter of a circle with a circumference of 5.4 meters. Round your answer to the nearest tenth. Use $\pi = 3.14$.

$$
\begin{array}{r}
1.71 \\
3.14\overline{)5.40.00} \\
3\ 14 \\
\hline
2\ 26\ 0 \\
2\ 19\ 8 \\
\hline
6\ 20 \\
3\ 14
\end{array}
$$

1.71 rounded to the nearest tenth is 1.7. The diameter is 1.7 meters.

To change a fraction to a decimal, follow these steps:

Step 1 Divide the numerator by the denominator.

Step 2 Put a decimal point to the right of the dividend. Add zeros as needed.

Step 3 Put the answer decimal point straight above the decimal point in the dividend.

Example

F. You have been given two values for π: $3\frac{1}{7}$ and 3.14. Change the mixed number to a decimal to show that they are the same. Round your answer to hundredths.

$$3\frac{1}{7} = \frac{22}{7}$$

$$
\begin{array}{r}
3.142 \\
7\overline{)22.00} \\
21 \\
\hline
1\ 0 \\
7 \\
\hline
30 \\
28 \\
\hline
20 \\
14 \\
\hline
6
\end{array}
$$

3.142 rounded to the nearest hundredth is 3.14.

Solve.

1. Find the perimeter of this figure.

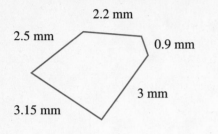

2. Find the missing dimension.

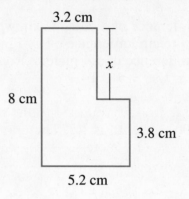

_____

_____

3. Find the area of this figure. Round your answer to the nearest hundredth.

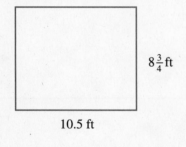

_____

4. An 8-foot long board is to be cut into pieces that are each 2.5 feet long. How many pieces can be cut from this board? How much is left?

5. The circumference of a circle is 32.8 meters. Find the diameter of the circle. Round your answer to the nearest tenth.

_____

_____

Operations With Ratios and Proportions

A **ratio** is a comparison of two numbers by division. Ratios can be written three ways: **as a fraction** ($\frac{1}{3}$), with the word **to** (1 to 3), or with a **colon** (1:3). Like fractions, ratios should be reduced to lowest terms. However, if the denominator is one, leave that number. A ratio needs two numbers.

Examples

A. To make a certain shade of pink paint, 4 gallons of white paint are mixed with 6 gallons of red paint. What is the ratio of white paint to red paint?

$\frac{2}{3}$ or   2 to 3   or   2:3

The first item listed is written first, or in the numerator.

B. Sandy has 12 red marbles and 4 green marbles. What is the ratio of red marbles to green marbles?

$\frac{12}{4}$ or $\frac{3}{1}$ or   3 to 1   or   3:1

A **proportion** is a statement that two ratios are equal. The same ratio or comparison can be expressed by each of these fractions: $\frac{1}{2}\ \frac{2}{4}\ \frac{3}{6}\ \frac{4}{8}\ \frac{5}{10}\ \frac{6}{12}\ \frac{8}{16}$. Any two of them form a proportion:

$\frac{1}{2} = \frac{2}{4}$ $\frac{2}{4} = \frac{6}{12}$ $\frac{8}{16} = \frac{3}{6}$

In any proportion, the product of the outer terms, **extremes,** equals the product of the inner terms, **means.**

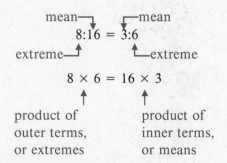

Therefore, $8 \times 6 = 48$, and $16 \times 3 = 48$.

If a proportion is written in fractional form, the cross products are equal.

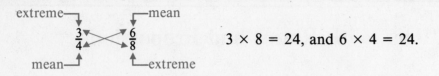

$$3 \times 8 = 24, \text{ and } 6 \times 4 = 24.$$

Proportions are used to convert measurements from one unit to another. They are also used to solve problems involving similar triangles and other types of geometric figures.

Examples

C. Change 40 inches to feet.

12 inches = 1 foot

Let x = the number of feet.

$\frac{12 \text{ in.}}{1 \text{ ft}} = \frac{40 \text{ in.}}{x \text{ ft}}$ Cross multiply.

$12(x) = 40(1)$

$x = \frac{40}{12}$

$x = 3\frac{1}{3} \text{ ft}$

Thus, 40 inches is $3\frac{1}{3}$ feet.

Geometric figures are said to be similar if they have the same shape. Notice that the angles in triangle *ABC* are the same measure as the angles in triangle *DEF*.

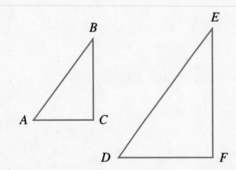

Triangle *ABC* is similar to triangle *DEF*. Corresponding sides of similar triangles are proportional.

$$\frac{AB}{DE} = \frac{BC}{EF} = \frac{CA}{FD}$$

D. Triangle *ABC* is similar to triangle *DEF*. Find the value of *x*.

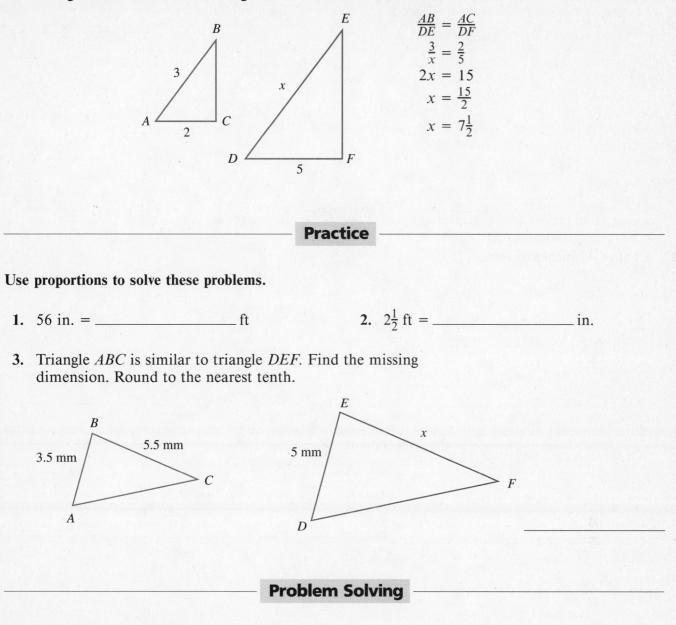

$$\frac{AB}{DE} = \frac{AC}{DF}$$

$$\frac{3}{x} = \frac{2}{5}$$

$$2x = 15$$

$$x = \frac{15}{2}$$

$$x = 7\frac{1}{2}$$

───────────────── **Practice** ─────────────────

Use proportions to solve these problems.

1. 56 in. = _____ ft

2. $2\frac{1}{2}$ ft = _____ in.

3. Triangle *ABC* is similar to triangle *DEF*. Find the missing dimension. Round to the nearest tenth.

_____

───────────────── **Problem Solving** ─────────────────

Solve the following problems.

4. The scale on a road map is 1 inch = 25 miles. If the distance between two towns measures $2\frac{1}{4}$ inches, about how many miles apart are the towns?

_____

5. The scale on a road map is 1 inch = 80 miles. If the distance between two towns measures $1\frac{3}{4}$ inches, about how many miles apart are the towns?

_____

21

6. A machinist made a gear $10\frac{1}{2}$ inches in diameter with 54 teeth. Using this same ratio, how many teeth will be on a 7-inch diameter gear?

_____

7. One kilometer is about .62 miles. About how many kilometers are in 50 miles? Round to the nearest tenth.

_____

8. A photograph 3 inches wide and 4 inches long is enlarged to 12 inches long. How wide is the enlargement?

_____

9. To make a certain shade of green, the ratio of black paint to green paint is 2 to 9. How many gallons of black paint should be mixed with 6 gallons of green paint to make this color?

_____

10. How much paint is needed for 3,500 square feet if one gallon covers 750 square feet? Round to the nearest gallon.

_____

Exponents and Roots

An **exponent** tells how many times a number has been multiplied by itself. Thus, the base number repeats itself in multiplication.

$$7^2 = \underbrace{7 \times 7}_{2 \text{ times}} = 49$$

exponent

Read "7 to the second power is 49" or "7 squared is 49."

$$5^3 = \underbrace{5 \times 5 \times 5}_{3 \text{ times}} = 125$$

Read "5 to the third power is 125" or "5 cubed is 125."

If there is no exponent following a number, the exponent is 1.

$$9 = 9^1 = 9$$

Any number with an exponent of zero equals 1.

$$100^0 = 1$$

Your calculator may have keys that will help you simplify problems involving exponents. Use the $\boxed{x^2}$ key to find the square of a number. To find the value of 5^2, do the following:

Step 1 Press $\boxed{5}$.

Step 2 Press $\boxed{x^2}$.

Step 3 Press $\boxed{=}$ and read the display. $5^2 = 25$

Use the $\boxed{y^x}$ key to find other powers. To find the value of 3^4, do the following:

Step 1 Press $\boxed{3}$.

Step 2 Press $\boxed{y^x}$.

Step 3 Press $\boxed{4}$.

Step 4 Press $\boxed{=}$ and read the display. $3^4 = 81$

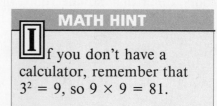

MATH HINT

If you don't have a calculator, remember that $3^2 = 9$, so $9 \times 9 = 81$.

A. Find the area of a circle with a radius of 3 cm. Use this formula:
$A = \pi r^2$

Substitute 3.14 for π and 3 for r in the formula:

$A = 3.14(3)(3)$ Note that r^2 means r times r, not r times 2.
    $= 28.26 \text{ cm}^2$ Simplify.

To find the **square root** of a number means to find the number that was multiplied by itself to get the number. The symbol used to show this operation is a **radical sign** $\sqrt{}$.

$\sqrt{36} = 6$ since
             $6^2 = 6 \times 6 = 36.$

The number 36 is an example of a **perfect square.** This means that the square root is a whole number. The square root of a number that is not a perfect square can be found by several methods. One way is by reading a table of square roots like the one found on page 25.

You can also use your calculator to help you simplify problems involving any roots. Use the $\boxed{\sqrt{}}$ key to find the square root of a number. To find $\sqrt{25}$, do the following:

Step 1   Press $\boxed{2}$, $\boxed{5}$.

Step 2   Press $\boxed{\sqrt{}}$.

Step 3   Read the display. $\sqrt{25} = 5$

On a calculator, use the $\boxed{\text{2ndF}}$ key and the $\boxed{y^x}$ key to find the roots, other than square roots, of numbers. To find $\sqrt[3]{32}$, read "the cube root of 32," do the following:

Step 1   Press $\boxed{3}$, $\boxed{2}$.

Step 2   Press $\boxed{\text{2ndF}}$ and $\boxed{y^x}$.

Step 3   Press $\boxed{3}$.

Step 4   Press $\boxed{=}$ and read the display.

      $\sqrt[3]{32} = 3.174802104 \approx 3.17$

A portion of a table of square roots is shown below.

n	√n	n	√n	n	√n	n	√n	n	√n	n	√n
1	1.000 000	51	7.141 428	101	10.04988	151	12.28821	201	14.17745	251	15.84298
2	1.414 214	52	7.211 103	102	10.09950	152	12.32883	202	14.21267	252	15.87451
3	1.732 051	53	7.280 110	103	10.14889	153	12.36932	203	14.24781	253	15.90597
4	2.000 000	54	7.348 469	104	10.19804	154	12.40967	204	14.28286	254	15.93738
5	2.236 068	55	7.416 198	105	10.24695	155	12.44990	205	14.31782	255	15.96872
6	2.449 490	56	7.483 315	106	10.29563	156	12.49000	206	14.35270	256	16.00000
7	2.645 751	57	7.549 834	107	10.34408	157	12.52996	207	14.38749	257	16.03122
8	2.828 427	58	7.615 773	108	10.39230	158	12.56981	208	14.42221	258	16.06238
9	3.000 000	59	7.681 146	109	10.44031	159	12.60952	209	14.45683	259	16.09348
10	3.162 278	60	7.745 967	110	10.48809	160	12.64911	210	14.49138	260	16.12452
11	3.316 625	61	7.810 250	111	10.53565	161	12.68858	211	14.52584	261	16.15549
12	3.464 102	62	7.874 008	112	10.58301	162	12.72792	212	14.56022	262	16.18641
13	3.605 551	63	7.937 254	113	10.63015	163	12.76715	213	14.59452	263	16.21727
14	3.741 657	64	8.000 000	114	10.67708	164	12.80625	214	14.62874	264	16.24808
15	3.872 983	65	8.062 258	115	10.72381	165	12.84523	215	14.66288	265	16.27882
16	4.000 000	66	8.124 038	116	10.77033	166	12.88410	216	14.69694	266	16.30951
17	4.123 106	67	8.185 353	117	10.81665	167	12.92285	217	14.73092	267	16.34013
18	4.242 641	68	8.246 211	118	10.86278	168	12.96148	218	14.76482	268	16.37071
19	4.358 899	69	8.306 624	119	10.90871	169	13.00000	219	14.79865	269	16.40122
20	4.472 136	70	8.366 600	120	10.95445	170	13.03840	220	14.83240	270	16.43168
21	4.582 576	71	8.426 150	121	11.00000	171	13.07670	221	14.86607	271	16.46208
22	4.690 416	72	8.485 281	122	11.04536	172	13.11488	222	14.89966	272	16.49242
23	4.795 832	73	8.544 004	123	11.09054	173	13.15295	223	14.93318	273	16.52271
24	4.898 979	74	8.602 325	124	11.13553	174	13.19091	224	14.96663	274	16.55295
25	5.000 000	75	8.660 254	125	11.18034	175	13.22876	225	15.00000	275	16.58312
26	5.099 020	76	8.717 798	126	11.22497	176	13.26650	226	15.03330	276	16.61325
27	5.196 152	77	8.774 964	127	11.26943	177	13.30413	227	15.06652	277	16.64332
28	5.291 503	78	8.831 761	128	11.31371	178	13.34166	228	15.09967	278	16.67333
29	5.385 165	79	8.888 194	129	11.35782	179	13.37909	229	15.13275	279	16.70329
30	5.477 226	80	8.944 272	130	11.40175	180	13.41641	230	15.16575	280	16.73320
31	5.567 764	81	9.000 000	131	11.44552	181	13.45362	231	15.19868	281	16.76305
32	5.656 854	82	9.055 385	132	11.48913	182	13.49074	232	15.23155	282	16.79286
33	5.744 563	83	9.110 434	133	11.53256	183	13.52775	233	15.26434	283	16.82260
34	5.830 952	84	9.165 151	134	11.57584	184	13.56466	234	15.29706	284	16.85230
35	5.916 080	85	9.219 544	135	11.61895	185	13.60147	235	15.32971	285	16.88194
36	6.000 000	86	9.273 618	136	11.66190	186	13.63818	236	15.36229	286	16.91153
37	6.082 763	87	9.327 379	137	11.70470	187	13.67479	237	15.39480	287	16.94107
38	6.164 414	88	9.380 832	138	11.74734	188	13.71131	238	15.42725	288	16.97056
39	6.244 998	89	9.433 981	139	11.78983	189	13.74773	239	15.45962	289	17.00000
40	6.324 555	90	9.486 833	140	11.83216	190	13.78405	240	15.49193	290	17.02939
41	6.403 124	91	9.539 392	141	11.87434	191	13.82027	241	15.52417	291	17.05872
42	6.480 741	92	9.591 663	142	11.91638	192	13.85641	242	15.55635	292	17.08801
43	6.557 439	93	9.643 651	143	11.95826	193	13.89244	243	15.58846	293	17.11724
44	6.633 250	94	9.695 360	144	12.00000	194	13.92839	244	15.62050	294	17.14643
45	6.708 204	95	9.746 794	145	12.04159	195	13.96424	245	15.65248	295	17.17556
46	6.782 330	96	9.797 959	146	12.08305	196	14.00000	246	15.68439	296	17.20465
47	6.855 655	97	9.848 858	147	12.12436	197	14.03567	247	15.71623	297	17.23369
48	6.928 203	98	9.899 495	148	12.16553	198	14.07125	248	15.74802	298	17.26268
49	7.000 000	99	9.949 874	149	12.20656	199	14.10674	249	15.77973	299	17.29162
50	7.071 068	100	10.00000	150	12.24745	200	14.14214	250	15.81139	300	17.32051

To determine the square root of 10, find 10 in the column headed n.
Read the number that is on the same line but under the column headed
√n. So, √10 is 3.162 (rounded to the nearest thousandth).

B. The area of a circle is 28 in.^{2} Find the radius. Use the formula $A = \pi r^2$.

$$28 = 3.14(r^2)$$
$$8.917 = r^2$$
$$\sqrt{8.917} = r$$
$$2.99 \text{ in.} = r$$

C. A right triangle has three sides. The two short sides are called the legs of the triangle. The longest side, which is opposite the right angle, is called the hypotenuse. This formula shows the relationship between the legs and the hypotenuse:

$$a^2 + b^2 = c^2$$

You can use this formula, called the Pythagorean Theorem, to find the length of the hypotenuse. You can also use this formula to find the length of a leg if the lengths of the hypotenuse and one leg are known.

To determine the value of c, find the square root of 25.

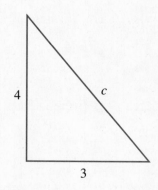

$$a^2 + b^2 = c^2$$
$$3^2 + 4^2 = c^2$$
$$9 + 16 = c^2$$
$$25 = c^2$$
$$5 = c$$

Find the values. Round to the nearest ten-thousandths.

1. 1^5 _____

2. 27^0 _____

3. 2^5 _____

4. $\sqrt{1}$ _____

5. $\sqrt{225}$ _____

6. $\sqrt{3}$ _____

7. $\sqrt{6}$ _____

8. 3^3 _____

9. 6^4 _____

Solve the following problems.

10. What is the square of 15? _____

11. The radius of a circle is 15 mm. Find the area of the circle. _____

12. The area of a circle is 21.5 m². Find the radius of the circle. _____

13. The legs of a right triangle measure 6 inches and 8 inches. Find the length of the hypotenuse. _____

14. The hypotenuse of a right triangle is 8 mm. One leg is 6 mm. Use the Pythagorean Theorem to find the length of the other leg. _____

Find the missing side of each of the following triangles. Show your work.

15.

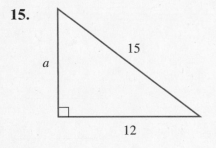

_____

16.

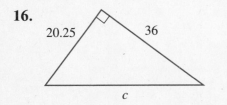

_____

Problem Solving: Identifying Key Words

Word problems are not difficult to master if you follow these steps:

Step 1   Read the problem and underline the key words. These words will generally relate to some mathematical reasoning.

Step 2   Make a plan to solve the problem. Ask yourself, Should I add, subtract, multiply, divide, round, or compare? You may have to do more than one of these operations for the same problem. Try to estimate what your answer will be. Will it be larger or smaller than the given number?

Step 3   Find the solution. Use your math knowledge to find your answer.

Step 4   Check the answer. Ask yourself, Is the answer reasonable? Is the answer near your estimate? Did you find what you were asked for?

--- **Example** ---

Paper towels sell for 3 rolls for $2.00. How much does one roll cost?

Step 1   The key words are **How much does one cost?** The answer must be smaller than $2.00. You are moving from a larger amount (3 rolls) to a smaller amount (1 roll).

Step 2   If you know what 3 rolls cost, you can find what 1 roll costs by setting up a ratio.

$$\frac{\$2.00}{3 \text{ rolls}} = \frac{x}{1 \text{ roll}}$$

Step 3   Solve the proportion.

$$\frac{\$2.00}{3} = \frac{x}{1}$$
$$3x = \$2.00$$
$$\frac{3x}{3} = \frac{\$2.00}{3}$$
$$x = \$0.6666$$

Rounded to the nearest cent, each roll costs $0.67.

Step 4   Check by adding the cost of 3 rolls.

$$\$0.67 + \$0.67 + \$0.67 = \$2.01$$

Notice that the sum is $0.01 more than $2.00 because the amount was rounded. Also, the estimate stated that the answer must be less than $2.00. The amount $0.67 is less than $2.00.

Solve the following problems.

1. Five boxes of macaroni and cheese cost $1.00. How much do 12 boxes cost? _____

2. A washing machine costs $459.00. It goes on sale for $399.00. How much money is saved? _____

3. A carpet costs $12.29 per yard. How much will $9\frac{1}{2}$ yards cost? _____

4. A vinyl floor tile costs $7.99 per square yard. How much will 12 square yards cost? _____

5. George bought 3 boxes of cereal that were on sale for $2.59 per box. He also bought 3 pounds of ground beef that sold for $1.89 per pound. If George gave the clerk a $20 bill for his groceries, how much change should he receive from the clerk? _____

6. The area of a room is 108 square feet. How many square yards of carpet will Molly need in order to carpet the room? If the carpet is on sale for $9.59 per square yard, how much will Molly pay for the carpet? _____

Posttest

Circle the correct answer for each question.

1. How many pieces of rope $\frac{5}{6}$ foot long can be cut from a 12-foot long rope?
   - **(1)** 10 pieces
   - **(2)** $\frac{5}{72}$ pieces
   - **(3)** 14 pieces
   - **(4)** $11\frac{5}{6}$ pieces
   - **(5)** $11\frac{1}{6}$ pieces

2. Triangle *ABC* is similar to triangle *DEF*. The dimensions are shown on the figures below. Find the length of *x*.

   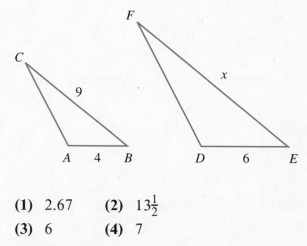

   - **(1)** 2.67
   - **(2)** $13\frac{1}{2}$
   - **(3)** 6
   - **(4)** 7
   - **(5)** More information is needed.

3. An electrician needs $20\frac{3}{4}$ feet of wire for one job and $32\frac{3}{8}$ feet for another job. If he has a roll of wire that contains 50 feet, how much more wire does he need?
   - **(1)** $3\frac{1}{8}$ feet
   - **(2)** $53\frac{1}{8}$ feet
   - **(3)** $2\frac{1}{2}$ feet
   - **(4)** $103\frac{1}{8}$ feet
   - **(5)** $2\frac{1}{8}$ feet

4. The legs of a right triangle measure 4 feet and 10 feet. Use the Pythagorean Theorem to find the length of the hypotenuse. $(a^2 + b^2 = c^2)$
   - **(1)** 14 feet
   - **(2)** 3.74 feet
   - **(3)** 116 feet
   - **(4)** 10.77 feet
   - **(5)** 13,456 feet

5. Susan wants to make macrame flowerpot hangers. If each hanger takes $12\frac{1}{2}$ feet of jute, how many inches of jute are needed for each hanger?

   (1) 1.04 in. (2) $24\frac{1}{2}$ in.

   (3) 150 in. (4) $144\frac{1}{2}$ in.

   (5) $1\frac{1}{2}$ in.

6. Use $F = \frac{9}{5}C + 32$ to change 45°C to a Fahrenheit reading.

   (1) 81°F (2) $42\frac{7}{9}$°F

   (3) 57°F (4) $138\frac{3}{5}$°F

   (5) 113°F

7. The area of a trapezoid is found by using the formula: $A = \frac{1}{2}(b_1 + b_2)h$. Find the area of a trapezoid with bases of 7 cm and 13 cm and a height of 9 cm.

   (1) $118\frac{1}{2}$ cm² (2) 180 cm²

   (3) 90 cm² (4) 360 cm²

   (5) 131 cm²

8. Find the perimeter of a rectangle with a length of 110 feet and a width of 95 feet. Use $P = 2l + 2w$.

   (1) 205 feet (2) 410 feet

   (3) $102\frac{1}{2}$ feet (4) 820 feet

   (5) 315 feet

9. Find the area of the rectangle described in Problem 8. Use $A = lw$.

   (1) 205 ft² (2) 1,045 ft²

   (3) 1,450 ft² (4) 410 ft²

   (5) 10,450 ft²

10. The circumference of a circle is given by the formula $C = \pi d$. Find the circumference of a circle with a diameter of 3 inches. Use $\pi = 3.14$.

   (1) 9.42 inches (2) 0.14 inch

   (3) 6.14 inches (4) 9.14 inches

   (5) 0.96 inch

Angles

Pretest

Answer each of the following questions.

1. One complementary angle measures 72°.
    What is the measure of the other angle?

    _____

2. What is an angle that measures more than
    90° and less than 180° called?

    _____

3. How are vertical angles related?

    _____

4. What type of angle is formed when two
    lines are perpendicular?

    _____

Use the figure below to answer questions 5–7. Circle the correct answer.

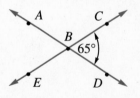

5. What is the measure of ∠*ABE*?

    (1) 25°   **(2)** 65°

    (3) 90°   **(4)** 115°

    (5) 180°

6. What is the measure of ∠*ABC*?

    (1) 25°   **(2)** 65°

    (3) 90°   **(4)** 115°

    (5) 180°

7. ∠*ABC* and ∠*ABE* are _____ .

   (1) complementary angles

   (2) supplementary angles

   (3) vertical angles

   (4) alternate interior angles

   (5) alternate exterior angles

8. What is the angle measure of a straight line?

_____

Use this figure to answer questions 9–12. Circle the correct answer.

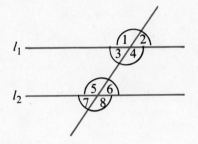

$l_1 \parallel l_2$

9. Name two alternate exterior angles.

   (1) ∠1 and ∠2   **(2)** ∠1 and ∠7

   (3) ∠7 and ∠8   **(4)** ∠2 and ∠8

   (5) ∠2 and ∠7

10. If ∠3 = 60°, ∠8 = _____ .

   (1) 30°   **(2)** 60°

   (3) 120°   **(4)** 150°

   (5) 240°

11. Name two alternate interior angles.

   (1) ∠3 and ∠6   **(2)** ∠1 and ∠2

   (3) ∠7 and ∠6   **(4)** ∠1 and ∠8

   (5) ∠4 and ∠6

12. ∠7 and ∠6 are _____ .

   (1) vertical angles

   (2) alternate interior angles

   (3) alternate exterior angles

   (4) corresponding angles

   (5) complementary angles

Introduction to Geometry

Geometry is the study of the size, shape, and position of objects in space.

Plane geometry is the study of flat surfaces. Flat surfaces such as a floor, a table top, or a chalkboard are examples of planes. A true plane extends endlessly in all four directions.

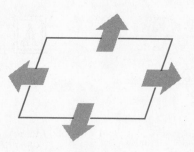

Some figures have only two dimensions, length and width. The figures below are plane figures.

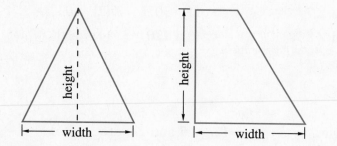

Solid geometry is the study of figures with three dimensions. Examples of figures with length, width, and height are a sugar cube and a box.

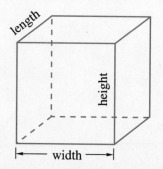

It is necessary to start your study of plane geometry with some definitions. In the chart below, a description and a symbol are given for each figure.

Figure	Description	Symbol
• A	**point:** a location in space; has no dimension	A
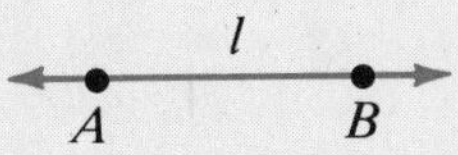 l ← A —— B →	**line:** two directions in space; extends endlessly in opposite directions	$\overleftrightarrow{AB}$ or line l
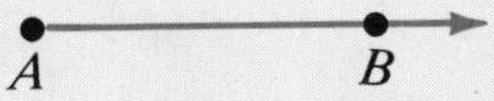 •A —— •B →	**ray:** a part of a line that starts at one point and extends endlessly in the other direction	$\overrightarrow{AB}$
 •A —————— •B	**line segment:** part of a line between two points; includes these two points	$\overline{AB}$ or $\overline{BA}$
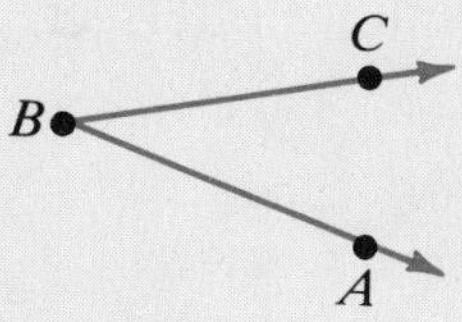 B •$\;$ •C → $\;$ •A →	**angle:** two rays intersecting at a common point; the common point is the vertex	$\angle ABC$ or $\angle CBA$ or $\angle B$

MATH HINT

|A| line is named by using any two points on the line or a lower-case letter.

MATH HINT

|A| ray is named by using the starting point called an endpoint and any other point on the ray.

MATH HINT

|A| line segment is named by using the endpoints of the segment.

MATH HINT

|A|n angle is named in several ways. You can use points on the rays and the common point. The vertex is the center letter. The vertex can be used alone to name the angle if it is clear which angle you mean.

In a plane, two lines can either meet or not meet. If the lines meet at a point, they are **intersecting** lines.

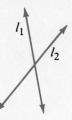

If two lines do not meet even if they are extended, they are **parallel** lines.

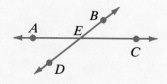

Practice

Use the figure below to answer questions 1–8.

1. Name four rays that have E as an endpoint.

    _____ _____

    _____ _____

2. Name the point where the lines intersect.

    _____

3. Name four angles.

    _____ _____

    _____ _____

4. Name six line segments.

    _____ _____

    _____ _____

    _____ _____

5. Name two lines.

_____   _____

6. Are $\overleftrightarrow{AC}$ and $\overleftrightarrow{DB}$ parallel lines or intersecting lines? Explain.

_____

_____

7. Are $\overrightarrow{AE}$ and $\overrightarrow{EA}$ the same ray? Explain.

_____

_____

8. Is $\angle AED$ and $\angle DEA$ the same angle? Explain.

_____

_____

Answer the following questions.

9. How many lines can be drawn through two points?

_____

10. Can a line be measured?

_____

Measuring and Drawing Angles

An **angle** is formed when two rays have the same starting point. The starting point becomes the vertex of the angle. The figure below shows $\overrightarrow{AC}$ and $\overrightarrow{AB}$ meeting at the starting point A.

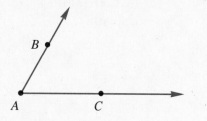

The symbol for angle is ∠, or ⦟. The angle can be named ∠*BAC,* ∠*CAB,* or ∠*A.*

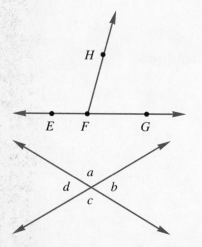

The figure at the left shows two angles:

  ∠*EFH*   ∠*GFH*

Notice that you cannot name either angle with only the letter *F.*

This figure shows four angles. Notice the angles are named by small letters inside the angles.

The opening between the rays determines the measure of the angle. Angles are measured in **degrees** (°).

You can use a **protractor** to measure angles. Marks on the protractor represent degrees.

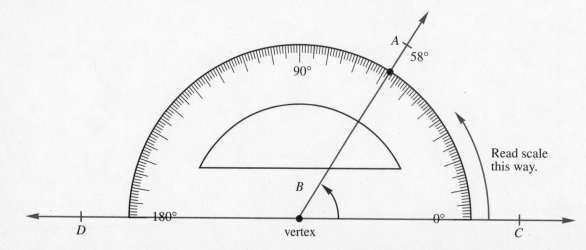

To measure $\angle CBA$, place the protractor along $\overrightarrow{BC}$ so that B, the vertex of the angle, is at the center of the protractor. Read the number on the scale where $\overrightarrow{BA}$ passes through the scale. (Make sure you read the scale that starts at 0° where $\overrightarrow{BC}$ passes through the scale.)

$\angle CBA$ measures 58°.

Angles are named by their measure.

Acute angles have a measure less than 90°.

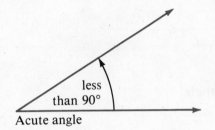

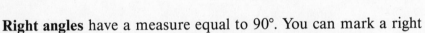

less than 90°

Acute angle

Right angles have a measure equal to 90°. You can mark a right angle by placing a box at the vertex of the angle.

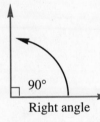

90°

Right angle

Obtuse angles have a measure greater than 90° and less than 180°.

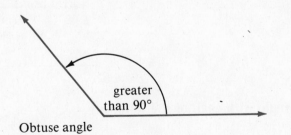

greater than 90°

Obtuse angle

A **straight line** has a measure equal to 180°.

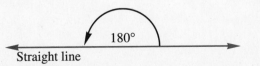

180°

Straight line

Find the measure of ∠ DCF. Name the angle.

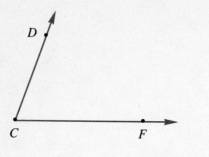

Use a protractor to measure the angle. The angle has a measure of 70°; therefore, it is an acute angle.

Use a protractor to measure each angle.

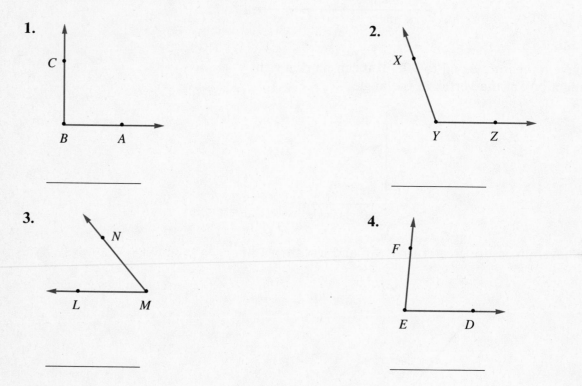

1.

_____

2.

_____

3.

_____

4.

_____

Use a protractor to draw an angle with each of the following measurements.

5. 73°

6. 125°

7. 90°   **8.** 10°

Use the figure below to answer questions 9 and 10.

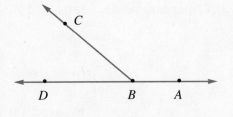

9. What kind of angle is ∠*ABC*?

_____

10. What kind of angle is ∠*DBC*?

_____

Problem Solving

---

11. What kind of angle is formed as the car turns the corner?

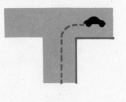

_____

12. A road is sloping on a curve to give drivers better control of their cars. What type of angle does the road use?

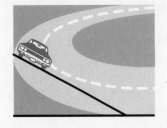

_____

Pairs of Angles

Certain pairs of angles are important in geometry. Three of these pairs are complementary angles, supplementary angles, and vertical angles.

If the sum of the measures of two angles is 90°, the angles are **complementary angles.** When complementary angles are placed next to each other, they make a right angle.

$50° + 40° = 90°$

Two angles with measurements of 40° and 50° are complementary angles because their total is 90°.

When the sum of the measures of two angles is 180°, the angles are **supplementary angles.** When supplementary angles are placed next to each other, they form a straight line.

$48° + 132° = 180°$

Two angles with measurements of 48° and 132° are supplementary angles because their total is 180°.

Examples

A. $\angle a$ and $\angle b$ are supplementary angles. $\angle a = 135°$. Find the measurement of $\angle b$.

Since $\angle a$ and $\angle b$ are supplementary angles,
$\angle a + \angle b = 180°$.

By substitution, $135° + \angle b = 180°$
$$\angle b = 180° - 135°$$
$$\angle b = 45°$$

Sometimes you need to use relationships between angles to solve a problem. You can use these relationships to write an equation and to solve for the measures of the angles.

B. An angle is 18° less than twice its complement. Find the measurement of the angle.

Let x = the measure of the angle.
Then, $90 - x$ = the measure of the complement of the angle.

So, the equation is $x = 2(90 - x) - 18$.

$$x = 180 - 2x - 18$$
$$x = 162 - 2x$$
$$3x = 162$$
$$x = 54$$

The measurement of the angle is 54°.

Vertical angles are formed when two lines intersect. Vertical angles are the angles opposite each other. Every pair of intersecting lines has two sets of vertical angles.

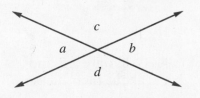

$\angle a$ and $\angle b$ are vertical angles. $\angle c$ and $\angle d$ are vertical angles.

Vertical angles have the same measure. If $\angle a$ has a measurement of 75°, then $\angle b$ has a measurement of 75°. If $\angle c$ has a measurement of 105°, then $\angle d$ has a measurement of 105°.

Examples

C. $\angle c$ and $\angle d$ are vertical angles. The measurement of $\angle c$ is 95°. Find the measurement of $\angle d$.

The measurement of $\angle d$ is 95°, because vertical angles have the same measure.

D. Use the figure below to find the measures of the two angles.

The two angles are vertical angles.

Therefore, $x = 3x - 38$
$$38 = 2x$$
$$19 = x$$

The vertical angles measure 19°.

Solve the following problems.

1. One complementary angle has a measurement of 35°. What is the measurement of the other angle?

   _____

2. One supplementary angle has a measurement of 125°. What is the measurement of the other angle?

   _____

Use the figure below to answer questions 3 and 4.

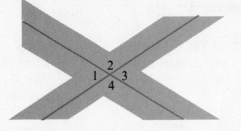

3. Two roads intersect as shown. What are ∠1 and ∠3 called? What is true about their measure?

   _____

   _____

4. What are ∠2 and ∠4 called? What is true about their measure?

   _____

   _____

For questions 5–8, solve for x.

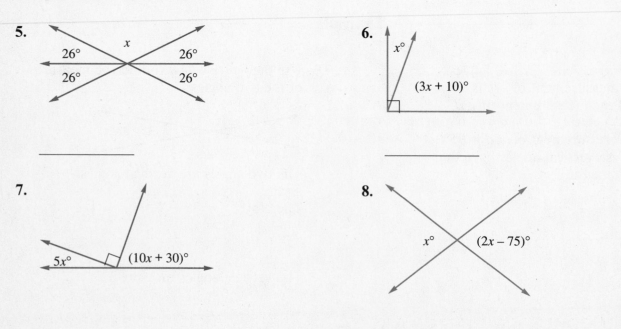

5.

   26° x 26°
   26°   26°

   _____

6.

   $x°$

   $(3x + 10)°$

   _____

7.

   $5x°$ $(10x + 30)°$

   _____

8.

   $x°$ $(2x - 75)°$

   _____

Miter Saw

A miter saw cuts boards at different angles. The saw has a scale showing angle measurement to the right and the left of 0°.

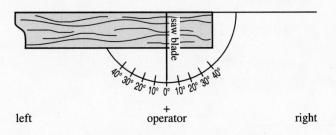

left   +   right
                    operator

To cut a board with a 60° angle, set the saw at 30° to the left (as you face the saw) of 0°. The piece of board cut off forms an angle of 30°. The angle of the board remaining is the complement, or 60°.

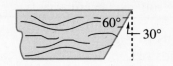

If you set the saw 30° to the right of 0°, the angle cut will be 90° + 30°, or 120°.

Find the setting of the miter saw needed to cut a piece of board with the following angles. Remember to tell whether the setting is to the left or the right of 0°.

1. 45°

_____

2. 30°

_____

3. 75°

_____

4. 112°

_____

Parallel Lines

Parallel lines are two lines in the same plane that never meet. These lines are always the same distance from each other.

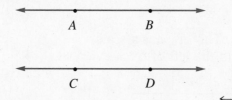

Lines AB and CD are parallel. This can be written as $\overleftrightarrow{AB} \parallel \overleftrightarrow{CD}$.

A line that intersects two parallel lines is a **transversal.** When a transversal cuts two parallel lines, several angles are formed.

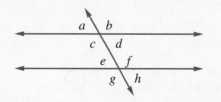

The angles formed and located in the same position compared to the transversal are **corresponding angles.** Corresponding angles are equal.

$\angle a$ corresponds to $\angle e$. $\angle c$ corresponds to $\angle g$.
$\angle b$ corresponds to $\angle f$. $\angle d$ corresponds to $\angle h$.

Angles on the outside of the parallel lines are **exterior angles,** and angles on the inside of the parallel lines are **interior angles.** Angles on opposite sides of the transversal are **alternate** angles.

You can identify alternate exterior angles. Alternate exterior angles are equal.

$\angle b$ and $\angle g$ form a pair of alternate exterior angles.
$\angle a$ and $\angle h$ form a pair of alternate exterior angles.

You can identify alternate interior angles when you have two parallel lines cut by a transversal.

$\angle d$ and $\angle e$ form a pair of alternate interior angles.
$\angle c$ and $\angle f$ form a pair of alternate interior angles.

Alternate interior angles are equal. You can show these pairs of angles are equal by using vertical angles and corresponding angles.

Find the measurement of each of the angles below. The two parallel lines are cut by a transversal. $\angle a = 120°$.

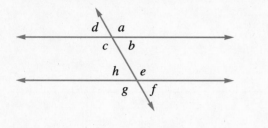

$\angle a = 120°$
$\angle a = \angle c = \angle e = \angle g$
So, $\angle c$, $\angle e$, and $\angle g$ all equal 120°.
$\angle a + \angle d = 180°$
So, $\angle d = 60°$
$\angle d = \angle b = \angle f = \angle h$
So, $\angle b$, $\angle f$, and $\angle h$ all equal 60°.

Practice

Use the figure below to answer the questions. The two parallel lines are cut by a transversal.

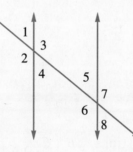

1. If $\angle 4 = 39°$, what are the measurements of $\angle 3$, $\angle 8$, and $\angle 7$?

   _____   _____

   _____

2. Name two angles with the same measure as $\angle 2$.

   _____   _____

3. Name two pairs of alternate interior angles.

   _____   _____

4. Name two pairs of corresponding angles.

   _____   _____

5. Name two pairs of alternate exterior angles.

   _____   _____

6. Name four pairs of vertical angles.

   _____   _____

   _____   _____

Perpendicular Lines

Two lines are **perpendicular** if they meet at right angles.

Example

You can write $\overleftrightarrow{AB} \perp \overleftrightarrow{CD}$. This is read "line AB is perpendicular to line CD."

When two parallel lines are cut by a transversal at a right angle, the measurements of all the angles are right angles. Look at the drawing in the Practice section.

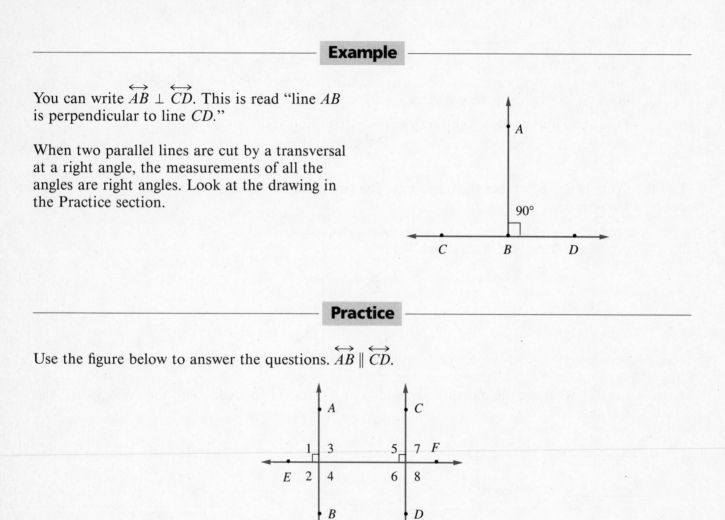

Practice

Use the figure below to answer the questions. $\overleftrightarrow{AB} \parallel \overleftrightarrow{CD}$.

1. If $\overleftrightarrow{EF} \perp \overleftrightarrow{AB}$ and $\overleftrightarrow{EF} \perp \overleftrightarrow{CD}$, what is the relationship between $\overleftrightarrow{AB}$ and $\overleftrightarrow{CD}$?

2. Name two pairs of alternate interior angles.

3. Name two pairs of alternate exterior angles.

4. Find the measurement of each angle.

Problem Solving: Read the Problem

The steps you have learned to solve word problems can be used with word problems that deal with geometric figures. Use the following steps:

Step 1   Read the problem and underline the key words. These words will generally relate to some mathematical reasoning.

Step 2   Make a plan to solve the problem. Ask yourself, Should I add, subtract, multiply, divide, round, or compare? You may have to do more than one of these operations for the same problem.

Step 3   Find the solution. Use your math knowledge to find your answer.

Step 4   Check your answer. Ask yourself, Is the answer reasonable? Did you find what you were asked for?

It is important to read the problem carefully when solving problems with geometric figures. Some key words in geometry problems include **parallel lines, perpendicular lines,** and **transversal.**

Example

A ladder resting against the side of a building forms an angle between its base and the ground measuring 75°. Find the complement and supplement of the angle formed.

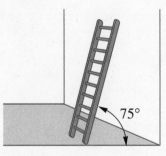

Step 1   Read the problem and identify the key words.
The key words are **complement** and **supplement.**

Step 2   Make a plan to solve the problem.
The sum of the complement of an angle and the angle is 90°.
The sum of the supplement of an angle and the angle is 180°.

Step 3   Find the solution.
Let x = the unknown angle. The complement of the angle is
$90 - x = 75$. So, the complement of $75°$ is $15°$. The supplement
of the angle is $180 - x = 75$. So, the supplement of $75°$ is $105°$.

Step 4   Check the answer.

$$75° + 15° = 90°$$
$$75° + 105° = 180°$$

Practice
---

1. A road crosses railroad tracks. The road
forms an angle of $50°$. Find the measures of
$\angle a$, $\angle b$, and $\angle c$.

_____

_____

_____

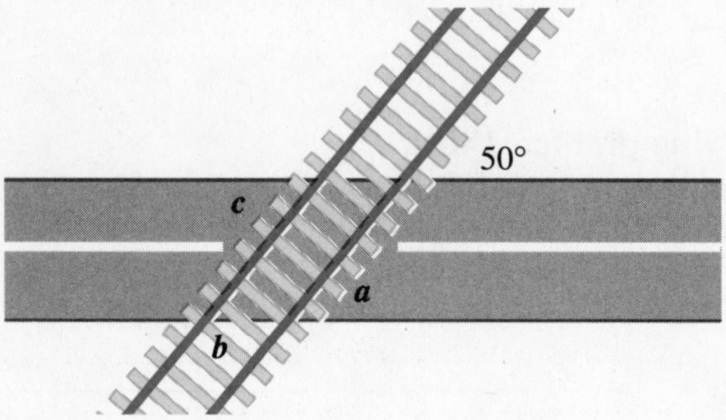

2. Two streets meet at an intersection. One
angle formed measures $60°$. Find the
measures of $\angle a$, $\angle b$, and $\angle c$.

_____

_____

_____

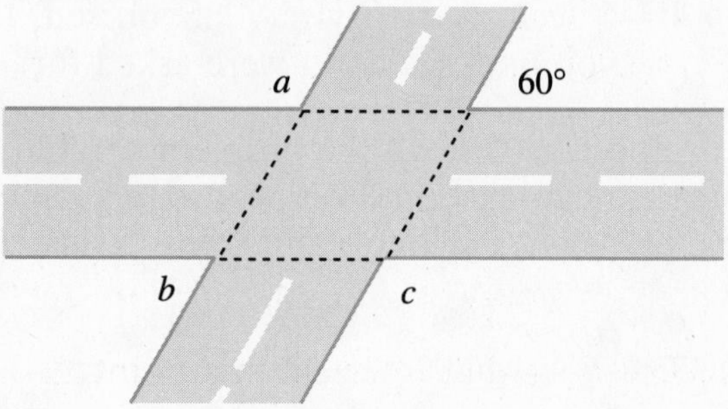

Circle the best answer for each question.

1. One complementary angle measures 88°. What is the measure of the other angle?

    (1) 2°   **(2)** 12°

    (3) 88°   **(4)** 90°

    (5) 92°

2. An angle that measures more than 90° and less than 180° is called _____ .

    (1) an acute angle

    (2) a right angle

    (3) an obtuse angle

    (4) a straight line

    (5) a complementary angle

3. Two lines that intersect in a point _____ .

    (1) are perpendicular   **(2)** are parallel

    (3) form vertical angles   **(4)** are equal

    (5) form complementary angles

4. If two lines are parallel, they _____ .

    (1) intersect in a right angle

    (2) are always the same distance apart

    (3) are equal

    (4) are perpendicular to each other

    (5) form vertical angles

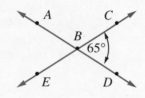

5. Name two rays formed by $\overleftrightarrow{AD}$.

    (1) $\overrightarrow{BA}$ and $\overrightarrow{BC}$ **(2)** $\overrightarrow{BA}$ and $\overrightarrow{AD}$

    (3) $\overrightarrow{BC}$ and $\overrightarrow{EC}$ **(4)** $\overrightarrow{BE}$ and $\overrightarrow{BC}$

    (5) $\overrightarrow{BA}$ and $\overrightarrow{EB}$

6. What is the measure of $\angle ABC$?

    (1) 25°   **(2)** 65°

    (3) 90°   **(4)** 115°

    (5) 180°

7. What is the measure of $\angle DBE$?

    (1) 25°   **(2)** 65°

    (3) 90°   **(4)** 115°

    (5) 180°

Use this figure to answer questions 8–10. $l_1 \parallel l_2$.

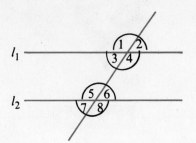

$l_1 \parallel l_2$

8. ∠2 and ∠6 are _____.
   (1) complementary angles
   (2) supplementary angles
   (3) corresponding angles
   (4) alternate interior angles
   (5) alternate exterior angles

9. Two alternate interior angles are _____.
   (1) ∠3 and ∠4   (2) ∠3 and ∠5
   (3) ∠3 and ∠6   (4) ∠5 and ∠6
   (5) ∠4 and ∠6

10. If ∠6 = 68°, ∠4 = _____.
    (1) 22°   (2) 68°
    (3) 112°   (4) 158°
    (5) 168°

U N I T

3

Triangles

———————————————————— **Pretest** ————————————————————

Circle the best answer for each question.

1. Which of the following can represent the measures of the three angles of a triangle?

   (1) 30°, 61°, 90°   **(2)** 30°, 40°, 110°

   (3) 45°, 55°, 90°   **(4)** 40°, 55°, 110°

   (5) 45°, 75°, 90°

2. In an isosceles triangle, the vertex angle is 55°. Find the measures of the other two angles.

   (1) 125°, 125°   **(2)** 55°, 70°

   (3) $62\frac{1}{2}°$, $62\frac{1}{2}°$ **(4)** 35°, 90°

   (5) $17\frac{1}{2}°$, $17\frac{1}{2}°$

3. What is the perimeter of this triangle?

   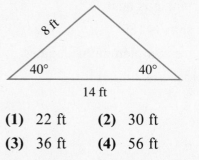

   (1) 22 ft   **(2)** 30 ft

   (3) 36 ft   **(4)** 56 ft

   (5) Insufficient information is given to solve this problem.

4. Find the area of this triangle.

   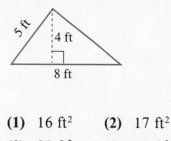

   (1) 16 ft²   **(2)** 17 ft²

   (3) 20 ft²   **(4)** 32 ft²

   (5) 64 ft²

5. Triangles *ABC* and *DEF* are similar. If $\overline{AB}$ is 4 in., $\overline{BC}$ is 6 in., and $\overline{DE}$ is 8 in., how long is $\overline{EF}$?

(1) 8 in. (2) 12 in.

(3) 24 in. (4) 36 in.

(5) 48 in.

6. If two sides and two angles of triangle *ABC* are equal to two sides and two angles of triangle *DEF*, _____ .

(1) $\triangle ABC \sim \triangle DEF$

(2) $\triangle ABC \cong \triangle DEF$

(3) the two triangles are the same

(4) the two triangles may be similar

(5) Insufficient information is given to solve this problem.

7. Find the area of a triangle with a base of 10 inches and a height of 12 inches.

(1) 22 in. (2) 60 sq in.

(3) 15.6 in. (4) 120 sq in.

(5) 484 sq in.

8. Find the distance (*x*) across the river.

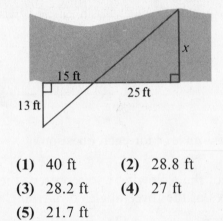

(1) 40 ft (2) 28.8 ft

(3) 28.2 ft (4) 27 ft

(5) 21.7 ft

9. Find *x*.

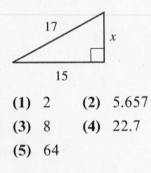

(1) 2 (2) 5.657

(3) 8 (4) 22.7

(5) 64

10. In any triangle, angles are equal if _____ .

(1) they are opposite equal sides

(2) the triangle is acute

(3) their sum is 180°

(4) they are complementary angles

(5) the sides are similar

Angles and Sides of Triangles

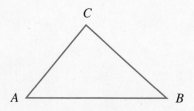

A **polygon** is a closed figure formed by line segments in a plane. The sides of a polygon do not cross each other.

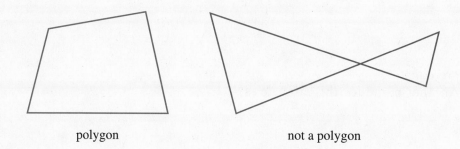

polygon   not a polygon

A **triangle** is a polygon with only three sides. Each point where the segments meet is a vertex. The sides are named by using the letters of the vertices at each end of the segment.

$\overline{AB}$, $\overline{BC}$, and $\overline{CA}$ are sides of $\triangle ABC$.

$\angle A$, $\angle B$, and $\angle C$ are angles of $\triangle ABC$.

Some triangles have special names. The names refer to special sides or special angles of the triangles.

Triangles With Special Sides

Scalene Triangle

All three sides of the triangle are of different lengths.

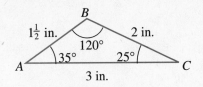

Isosceles Triangle

Two sides of the triangle are equal.
The angles opposite the two equal sides are equal.
These angles are called **base angles.**
The third angle is called the **vertex angle.**

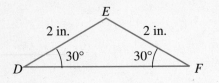

$\overline{DE}$ and $\overline{EF}$ are equal.
$\angle EDF$ and $\angle EFD$ are equal. These are called base angles.
$\angle DEF$ is the vertex angle.

Equilateral Triangle

All three sides of the triangle are equal.
The angles are all equal also. Another name for this triangle is equiangular.

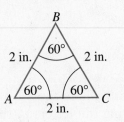

Triangles With Special Angles

Acute Triangle

Each angle of the triangle measures less than 90°.

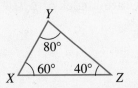

Obtuse Triangle

One angle of the triangle is greater than 90°.

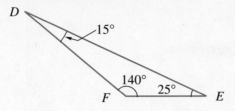

Right Triangle

One angle of the triangle measures 90°.

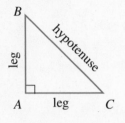

The side, $\overline{BC}$, opposite the 90° angle is called the hypotenuse. This side is the longest side. $\overline{AB}$ and $\overline{AC}$ are called legs.

The sum of the three angles in a triangle is always 180°.

Cut a triangle out of a sheet of paper. Label it and tear it as shown.

Place ∠A, ∠B, and ∠C together.

Place your protractor's zero on the common vertex of the angles. What is the total measure of the three angles? 180°

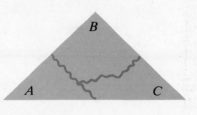

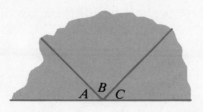

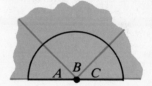

You can also show the sum of the angles of a triangle is 180° by using parallel lines. Draw ∆ABC so the base of the triangle, $\overline{AB}$, is parallel to a line drawn through the vertex C.

Alternate interior angles are formed. So, ∠1 and ∠4 are equal and ∠2 and ∠5 are equal.

∠3, ∠4, and ∠5 form a straight line. Therefore, ∠3 + ∠4 + ∠5 = 180°, and ∠3 + ∠1 + ∠2 = 180.

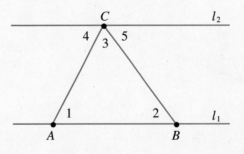

If you know the measures of two angles of a triangle, you can find the measure of the third angle. Subtract the sum of the two known angles from 180° to find the measurement of the third angle.

Examples

A. △ABC is an isosceles triangle with base angles of 50°. Find the measure of the third angle.

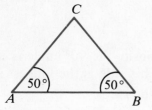

$\angle A = 50°$
$\angle B = 50°$
$\angle A + \angle B = 100°$
$180° - 100° = 80°$ Subtract the sum of $\angle A$ and $\angle B$ from 180°.

So, $\angle C = 80°$

B. Explain why an obtuse triangle cannot have more than one obtuse angle.

The total measure of the angles of a triangle is always 180°. An obtuse angle has a measure of more than 90°. The other two angles must have a sum less than 90°. Therefore, neither of these two angles can be an obtuse angle.

Practice

Identify whether the following triangles are scalene, isosceles, or equilateral. Then find the measurement of the missing angle.

1.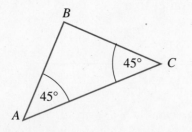

_____   ∠ B= _____

2.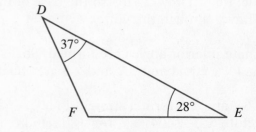

_____   ∠ F= _____

58

3.

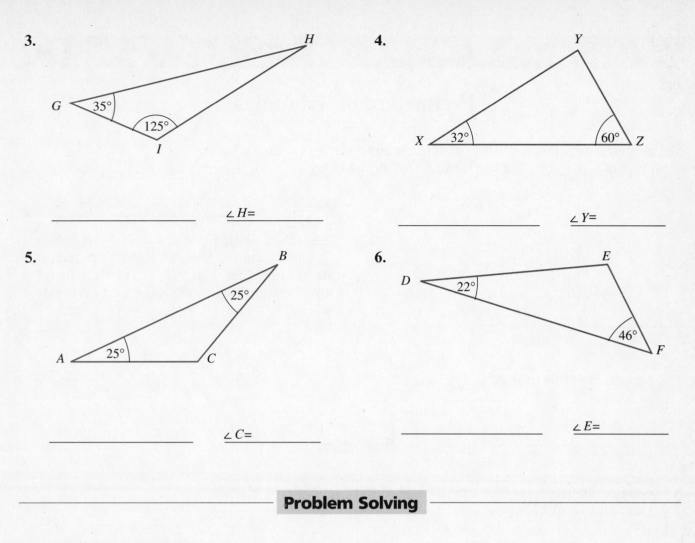

_____ ∠ H= _____

4.

∠ Y= _____

5.

_____ ∠ C= _____

6.

_____ ∠ E= _____

Problem Solving

Solve the following problems.

7. An equilateral triangle has three equal angles. What is the measure of each angle?

   _____

8. In an isosceles triangle, the vertex angle is 70°. Find the measures of the other two angles.

   _____

9. A right triangle contains a 30° angle. What is the measure of the third angle?

   _____

10. One angle of a right triangle is 90°. What is the sum of the other two angles?

   _____

Perimeters of Triangles

The **perimeter** of a polygon is the distance around the figure. To find the perimeter of a polygon, add the measures of all the sides.

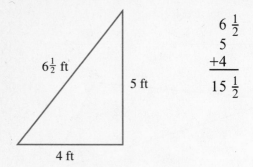

$$6\frac{1}{2}$$
$$5$$
$$+4$$
$$\overline{15\frac{1}{2}}$$

> **MATH HINT**
>
> **W**hen adding measurements, remember to add like units. You can add feet to feet, inches to inches, and yards to yards. Change all measurements to the same unit before adding.

The perimeter of this triangle is $15\frac{1}{2}$ feet.

Examples

A. What is the perimeter of a triangle with side measurements of 2 cm, 3 cm, and 18 mm?

First, change all the measurements to the same unit.

$$18 \text{ mm} = 1.8 \text{ cm}$$

Then, add the lengths.

$$2 + 3 + 1.8 = 6.8$$

The perimeter is 6.8 cm.

B. Each of the equal sides of an isosceles triangle is 8 inches long. The third side measures 4 inches. What is the perimeter of the triangle?

$$8 + 8 + 4 = 20$$

The perimeter is 20 inches.

C. The perimeter of a triangular park is 167 feet. One side measures 59 feet and a second side measures 38 feet. What is the measure of the third side?

The perimeter of a triangle is the sum of the measures of all three sides.

Let x = the measure of the third side.

$$59 + 38 + x = 167$$
$$x = 167 - 97$$
$$x = 70$$

The measure of the third side is 70 feet.

Practice

For questions 1–4, find the perimeter of each figure.

1.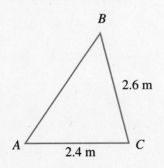

     3 in.   C

    B

   5 in.   7 in.

     A

_____

2. D

         8 yd

  5 yd 2 ft

   F   2 yd 2 ft   E

_____

3. H

  2 m 5 cm   2 m 5 cm

  G   K
     3 m 5 cm

_____

4. Y

  4 ft 2 in.   4 ft

  X   5 ft 10 in.   Z

_____

For questions 5 and 6, find the missing dimension.

5. Perimeter = 7.8m

      B

        2.6 m

  A   2.4 m   C

_____

6. Perimeter = 10 ft 4 in.

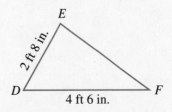

       E

  2 ft 8 in.

  D   4 ft 6 in.   F

_____

Solve the following problems.

7. Rafters for a garage measure 10 feet in length. The beams measure 14 feet. If two rafters and one beam are combined in the shape of a triangle, how much wood is required for each triangle?

_____

8. The perimeter of an equilateral triangle is 2 feet. Find the measure of each side in inches.

_____

9. The sum of any two sides of a triangle is greater than the third side. True or false. Explain your answer. (Draw a picture to illustrate your answer.)

_____

10. Complete the following statement:

In any triangle, the largest angle is opposite the longest side and the smallest angle is opposite the _____ side.

Areas of Triangles

The **area** of a polygon is the amount of surface it covers.

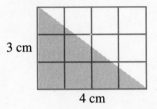

3 cm

4 cm

Look at the triangle. Count the number of squares to find the area of the triangle.
The area of this triangle is about 6 cm² (read "six square centimeters").

You can use the formula $A = \frac{1}{2}bh$ to find the exact answer, where A is the area, b is the base, and h is the height. The area of a triangle is equal to one-half the base times the height. So,

$A = \frac{1}{2}bh$

$A = \frac{1}{2} \cdot 4 \cdot 3$

$A = \frac{1}{2} \cdot 12$

$A = 6$

Example

Find the area of $\triangle ABC$.

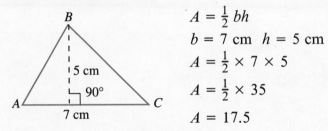

$A = \frac{1}{2}bh$

$b = 7$ cm $h = 5$ cm

$A = \frac{1}{2} \times 7 \times 5$

$A = \frac{1}{2} \times 35$

$A = 17.5$

The area of $\triangle ABC$ is 17.5 square centimeters.

MATH HINT

The measures of all the sides of a triangle are not used to find the area. Only the measures of the base and the height are used to find the area of a triangle.

Find the area of each triangle.

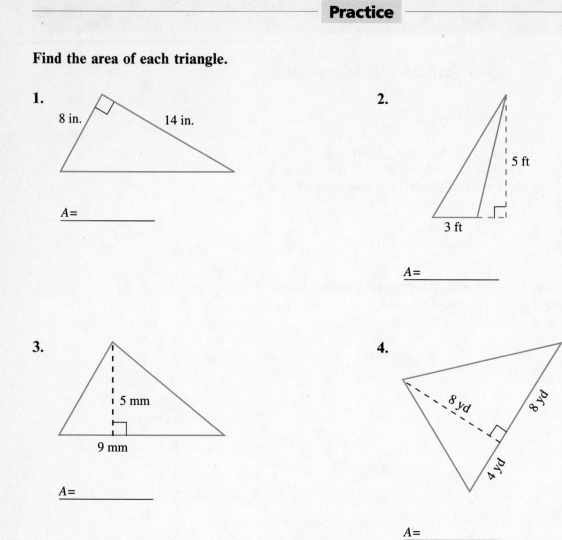

1.

8 in.   14 in.

A= _____

2.

5 ft

3 ft

A= _____

3.

5 mm

9 mm

A= _____

4.

8 yd

8 yd

4 yd

A= _____

Solve the following problem.

5. Angel is putting vinyl siding on his house. The gables (ends) of the house are triangular. Find the amount of siding needed to cover each gable.

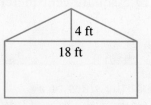

4 ft

18 ft

_____

Congruent Triangles

Two triangles are **congruent** if they are the same shape and size. Two triangles have the same shape if the measures of the angles of one triangle are equal to the measures of the angles of the other triangle. Two triangles have the same size if the measures of the sides of one triangle are equal to the measures of the sides of the other triangle.

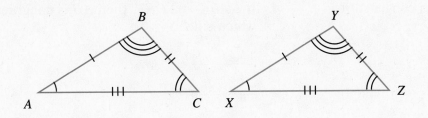

When referring to congruent triangles, list corresponding vertices in order. The symbol ≅ means "congruent to."

$$\triangle ABC \cong \triangle XYZ$$

Example

List the corresponding sides and angles of the congruent triangles *NOP* and *QRS*.

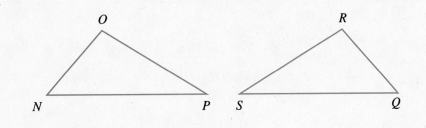

$$\angle N = \angle Q \quad \angle O = \angle R \quad \angle P = \angle S$$

$$\overline{NO} = \overline{QR} \quad \overline{OP} = \overline{RS} \quad \overline{NP} = \overline{QS}$$

Answer each of the following questions.

1. $\triangle ABC \cong \triangle DEF$. If $\angle B = 35°$, what is the measure of $\angle E$?

   _____

2. $\triangle LMN \cong \triangle XYZ$. If $LM = 10$, what is the measure of XY?

   _____

3. $\triangle ABC \approx \triangle DEF$, then $\overline{BC}$ is congruent to which side?

   _____

4. If $\triangle ABC \cong \triangle DEF$, then $\overline{AB}$ is congruent to which side?

   _____

Use the figures below to answer questions 5–8.

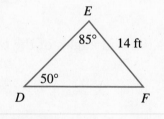

5. What is the measure of $\angle F$?
   (1) 85°
   (2) 50°
   (3) 45°
   (4) 35°
   (5) cannot be determined

6. $\overline{AC} =$ _____
   (1) DE
   (2) EF
   (3) $AB + BC$
   (4) DF
   (5) 95°

7. What is the measure of $\overline{DE}$?
   (1) 9 ft
   (2) 13 ft
   (3) 14 ft
   (4) 20 ft
   (5) cannot be determined

8. Which of these statements is correct?
   (1) $\triangle ABC \sim \triangle DEF$
   (2) $\triangle ABC \cong \triangle DEF$
   (3) $\triangle ACB \cong \triangle FDE$
   (4) $\triangle CAB \cong \triangle FED$
   (5) $\triangle BAC \cong \triangle EFD$

Similar Triangles

Two triangles are **similar** if they have the same shape.
Two triangles have the same shape if the measures of the angles of one triangle are equal to the measures of the angles of the other triangle.

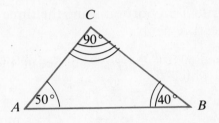 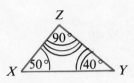

The angles with the same measure are corresponding angles.

$\angle A = \angle X$, so $\angle A$ corresponds to $\angle X$.
$\angle B = \angle Y$, so $\angle B$ corresponds to $\angle Y$.
$\angle C = \angle Z$, so $\angle C$ corresponds to $\angle Z$.

> **MATH HINT**
>
> **C**orresponding angles are shown on the triangles by curved lines. Angles with a matching number of curved lines correspond.

When referring to similar triangles, corresponding vertices are named in the same order.

Similar triangles may share parts. In $\triangle ABC$, $\overline{EF}$ is parallel to the base $\overline{AB}$.
The sides $\overline{AC}$ and $\overline{BC}$ are transversals passing through the parallel sides.

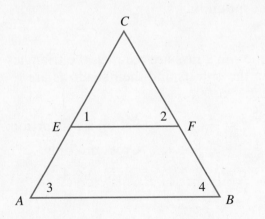

By corresponding angles, you can see the following:

$\angle 1$ is equal to $\angle 3$
$\angle 2$ is equal to $\angle 4$
$\angle ABC$ is equal to $\angle EFC$

Therefore, $\triangle ABC \sim \triangle EFC$.

The symbol $\sim$ means "is similar to."

Proportions can be used to find the length of a side of one of two similar triangles.

Example

A. $\triangle ABC \sim \triangle DEF$. Find the length of DF.
Let x = the length of side DF.

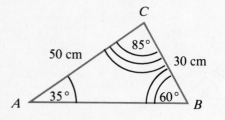

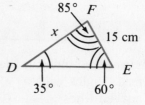

Write a proportion using the three known measurements.

Step 1 $\frac{AC}{DF} = \frac{BC}{EF}$ Set up the proportion.

            $\frac{50}{x} = \frac{30}{15}$

Step 2   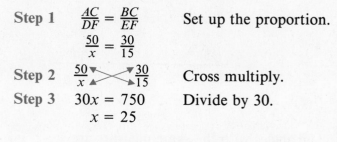   Cross multiply.

Step 3 $30x = 750$ Divide by 30.

           $x = 25$

The length of DF is 25 cm.

Similar triangles can be used to estimate heights and distances that would be hard to measure.

Example

B. A vertical yardstick casts a 4-foot shadow at the same time a building casts a 60-foot shadow. How tall is the building?

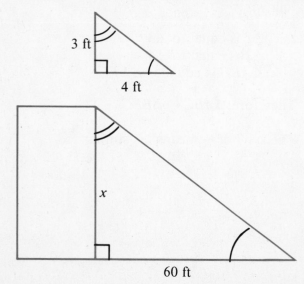

Because the sun's rays are parallel, the triangles formed by the objects and their shadows are similar.

Step 1 $\frac{3}{x} = \frac{4}{60}$ Set up a proportion.

Step 2 $\frac{3}{x} \diagdown \frac{4}{60}$ Cross multiply.

Step 3 $4x = 180$ Divide by 4.

         $x = 45$

The building is 45 feet tall.

C. A scout troop estimated the distance across a river by making the measurements shown below. How wide is the river?

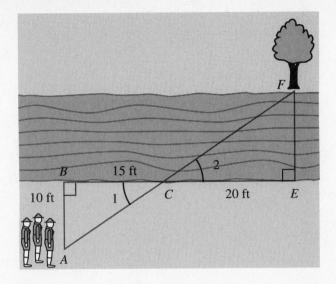

You know $\triangle ABC \sim \triangle FEC$ because the corresponding angles are equal.

Write a proportion using corresponding sides. Then substitute the known measurements. Let $r =$ the width of the river.

$$\frac{FE}{AB} = \frac{CE}{CB}$$
$$\frac{r}{10} = \frac{20}{15}$$
$$15r = 20(10)$$
$$15r = 200$$
$$r = 13\tfrac{1}{3}$$

The river is about $13\tfrac{1}{3}$ feet wide.

Practice

For questions 1 and 2, identify the proportions for the corresponding sides of the similar triangles.

1.

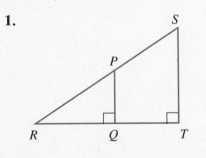

_____

2.

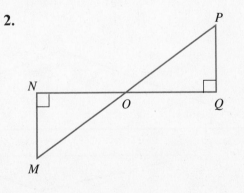

_____

69

Use this figure to answer questions 3 and 4.

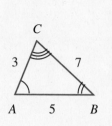

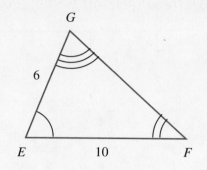

3. Martha and Jesse compared their solutions for the similar triangles shown above. They discovered that their proportions were not the same.

$\frac{3}{7} = \frac{6}{FG}$ $\frac{7}{3} = \frac{6}{FG}$

Martha's proportion   Jesse's proportion

Which proportion is correct?   _____

4. Solve both proportions for *FG*.

_____   _____

Problem Solving

Find the value of *x* in each pair of similar triangles. Show your work.

5.

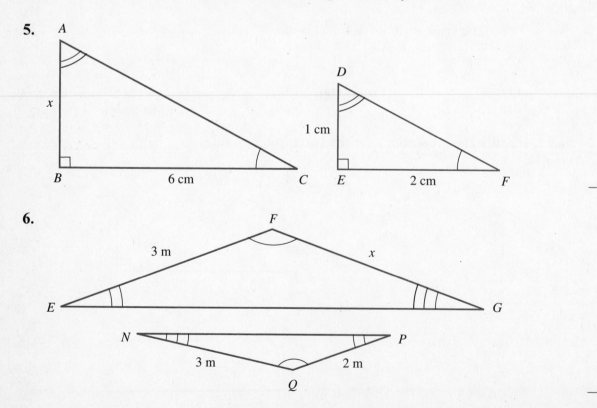

_____

6.

_____

70

Solve the following problems. Show your work.

7. △*ABE* ~ △*CDE*. Find *CD*.

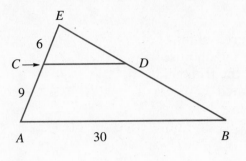

8. A skier made the measurements shown below. How wide is the jump?

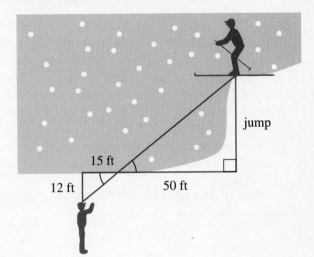

_____   _____

9. Mary Beth wants to estimate the height of a flagpole in order to replace the rope. She is $5\frac{1}{2}$ feet tall and casts a shadow of 5 feet when the flagpole casts a shadow of 65 feet. How tall is the flagpole?

10. Claudia Fischer, a telephone lineworker, is 6 feet tall. She casts a shadow 7 feet long at the same time a telephone pole casts a shadow 28 feet long. How tall is the pole?

_____   _____

Pythagorean Theorem

A **right triangle** is a triangle with one 90° angle. The two shorter sides of the triangle are called legs, and the longest side is called the hypotenuse. The legs are the sides of the right angle. The longest side of the triangle is opposite the right angle. The sides of a right triangle are related according to a special rule called the Pythagorean Theorem.

The **Pythagorean Theorem** states the following:

The sum of the squares of the legs of a right triangle equals the square of the hypotenuse.

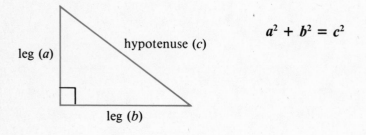

$$a^2 + b^2 = c^2$$

You can use the Pythagorean Theorem to find the measures of the sides of a right triangle.

Examples

A. Find the length of the hypotenuse (c).

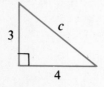

Step 1	$a^2 + b^2 = c^2$	Use the formula for the Pythagorean theorem.
	$3^2 + 4^2 = c^2$	Substitute the known values.
Step 2	$3 \cdot 3 + 4 \cdot 4 = c^2$	Simplify the powers.
	$9 + 16 = c^2$	
	$25 = c^2$	
Step 3	$\sqrt{25} = \sqrt{c^2}$	Solve for c by finding the
	$5 = c$	square root of both sides.

B. Find the length of the missing leg.

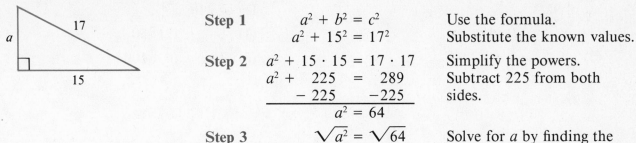

Step 1 $\qquad a^2 + b^2 = c^2 \qquad$ Use the formula.
$\qquad\qquad\quad a^2 + 15^2 = 17^2 \qquad$ Substitute the known values.

Step 2 $\quad a^2 + 15 \cdot 15 = 17 \cdot 17 \qquad$ Simplify the powers.
$\qquad\qquad\; a^2 + \;\;225 \;\;\;= \;\;289 \qquad$ Subtract 225 from both
$\qquad\qquad\qquad - 225 \qquad -225 \qquad$ sides.
$\qquad\qquad\qquad\qquad\quad a^2 = 64$

Step 3 $\qquad\qquad \sqrt{a^2} = \sqrt{64} \qquad$ Solve for a by finding the
$\qquad\qquad\qquad\quad a = 8 \qquad$ square root of both sides.

MATH HINT

Y̲ou can use either a table or a calculator to find the square or square root of a number. On a calculator, the x^2 key is used to find the square of a number. The $\sqrt{x}$ key is used to find the square root of a number.

Practice

The measure of 3 sides of a triangle are given. Determine if each triangle is a right triangle.

1. 7 ft, 9 ft, 6 ft _____ **2.** 5m, 12m, 13m _____

3. 9 in., 12 in., 14 in. _____ **4.** 4m, 7m, 5m _____

Problem Solving

Circle the best answer for each question.

5. Find the hypotenuse of a right triangle with legs measuring 6 and 8.
(1) 10   (2) 50
(3) 14   (4) 100

6. Find the hypotenuse of a right triangle with legs of 30 and 40.
(1) 2500   (2) 50
(3) 1250   (4) 7

7. Find the length of the other leg of a right triangle that has a hypotenuse of 15 and a leg of 9.

   (1) 6   **(2)** 18

   (3) 12   **(4)** 153

8. Find the length of the other leg of a right triangle that has a hypotenuse of 26 and a leg of 24.

   (1) 2   **(2)** 50

   (3) 10   **(4)** 100

9. Find the length of the diagonal of a square having sides measuring 2 inches.

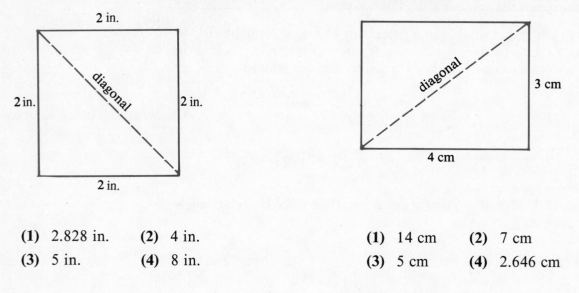

   (1) 2.828 in.   **(2)** 4 in.

   (3) 5 in.   **(4)** 8 in.

10. Find the length of the diagonal of a rectangle that measures 4 centimeters by 3 centimeters.

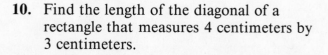

   (1) 14 cm   **(2)** 7 cm

   (3) 5 cm   **(4)** 2.646 cm

LIFE SKILL

Making a Support

To make a table, often a carpenter places a support under it. The length of the support is found by using the Pythagorean Theorem.

The tabletop and the table leg meet at a right angle. The support is the hypotenuse of the triangle. Use the theorem and substitute the lengths from the table.

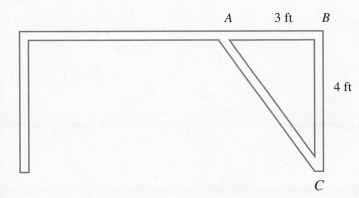

Let x^2 = the length of AC squared

$$3^2 + 4^2 = x^2$$
$$9 + 16 = x^2$$
$$25 = x^2$$
$$5 = x$$

The support needs to be 5 feet.

1. A carpenter needs to make a large platform with a support placed eight feet from the end of the table and placed six feet down on the leg. How long is the support to be cut?

   _____

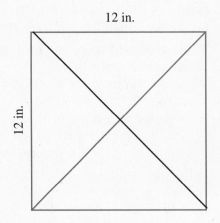

2. A frame for a picture is very flimsy. To make the frame stand more firm, supports need to be stapled at the four vertices as shown in the figure. How much wood does it take to make the supports?

   _____

Problem Solving: Make a Plan

The steps you have learned for solving word problems can be used with word problems that deal with geometric figures. Use the following steps:

Step 1   Read the problem and identify the key words. Underline the key words. These words will generally relate to some mathematical reasoning.

Step 2   Make a plan to solve the problem. Ask yourself, Should I add, subtract, multiply, divide, round, or compare? You may have to do more than one of these operations for the same problem.

Step 3   Find the solution. Use your math knowledge to find your answer.

Step 4   Check your answer. Ask yourself, Is the answer reasonable? Did you find what you were asked for?

It is important to make a plan to solve the problem. You should ask yourself several questions. What do I want to find? What do I know? How do I find what I need?

Example

A triangular garden in the park needs to have a fence placed around it. The lengths of the sides of the garden are 15 feet, 24 feet, and 45 feet. If the material for the fence costs $10.50 per foot, what will be the cost of materials for making the fence?

Step 1   Read the problem and identify the key words. The key words are **around, lengths,** and **how much.**

Step 2   Make a plan to solve the problem. What do I want to find?
The cost of the material.

What do I know? The length of each side, the cost of the material per foot.

How do I find what I need?
First, find the perimeter. Then, multiply the cost per foot by the number of feet in the perimeter.

Step 3   Solve the problem.

$$15 + 24 + 45 = 84$$

The perimeter is 84 feet.

$$84 \times \$10.50 = \$882.00$$

The cost of the materials is $882.

Step 4   Check your answer. Estimate the perimeter.

$$20 + 20 + 50 = 90$$

Estimate the cost.

$$90 \times \$10.50 = \$945$$

The estimate of the cost is $945. This estimate is close to the answer found above. The answer is reasonable.

Practice

1. Mary wants to paint the triangular tabletop shown below. She needs 1 quart of paint per 10 square feet. How much paint does she need?

    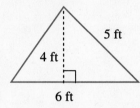

    _____

2. A tower needs a support wire placed in the ground 9 feet from the base. If the tower is 12 feet tall, how long is the support wire?

    _____

3. A bookcase needs supports at two of the corners. Each support is positioned 18 inches from the corner and 24 inches down the side. How much wood does it take for two supports?

    _____

    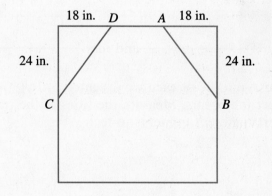

Work Triangles

According to home economists, a kitchen's efficiency can be measured by the distance between its three work centers: the refrigerator, the stove, and the sink.

The distance between the three work centers usually forms a **work triangle.** The perimeter of an efficient triangle should be no more than 22 feet.

Find the perimeter of the work triangle for this model kitchen.

stove   refrigerator

$\frac{5}{4}$ in.

$\frac{7}{4}$ in.

$\frac{11}{4}$ in.

sink

Mark the middle of each work center.
Connect the points. Measure the sides of the triangle ($\frac{1}{4}$ inch = 1 foot).
Convert the measurements to feet.

$$\frac{\frac{1}{4}\text{ inch}}{1\text{ foot}} = \frac{\frac{5}{4}\text{ inch}}{x\text{ feet}}$$

$$\frac{1}{4}x \cdot 4 = \frac{5}{4} \cdot 4$$

$$x = 5 \text{ feet}$$

$$\frac{5}{4} \text{ inches} = 5 \text{ feet}$$

$$\frac{\frac{1}{4}\text{ inch}}{1\text{ foot}} = \frac{\frac{7}{4}\text{ inch}}{y\text{ feet}}$$

$$\frac{1}{4}y \cdot 4 = \frac{7}{4} \cdot 4$$

$$y = 7 \text{ feet}$$

$$\frac{7}{4} \text{ inches} = 7 \text{ feet}$$

$$\frac{\frac{1}{4}\text{ inch}}{1\text{ foot}} = \frac{\frac{11}{4}\text{ inch}}{z\text{ feet}}$$

$$4 \cdot \frac{1}{4}z = \frac{11}{4} \cdot 4$$

$$z = 11 \text{ feet}$$

Add the sides:

     5 feet + 7 feet + 11 feet = 23 feet.

The perimeter of the work triangle is 23 feet. **This kitchen work area is not efficient.**

Find the perimeter of each work triangle $\left(\frac{1}{4} \text{ inch} = 1 \text{ foot}\right)$. Which kitchen is more efficient (requires less walking)?

1.

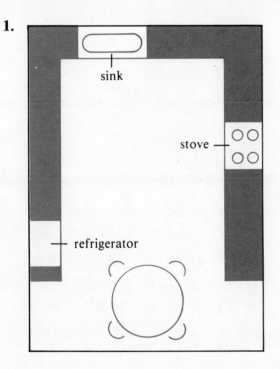

2.

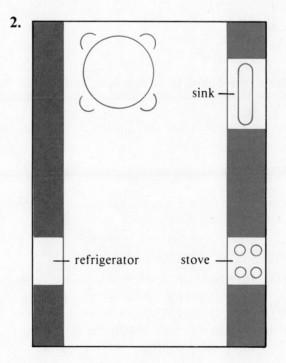

                _____

Posttest

Circle the best answer for each question.

1. If one angle of a triangle measures 45°, which of the following can represent the measures of the other two angles?

   (1) 35°, 90°   **(2)** 30°, 15°

   (3) 45°, 80°   **(4)** 40°, 130°

   (5) 45°, 90°

2. In an isosceles triangle, the vertex angle is 35°. Find the measures of the other two angles.

   (1) 55°, 90°   **(2)** 35°, 110°

   (3) $72\frac{1}{2}°$, $72\frac{1}{2}°$ **(4)** 145°, 145°

   (5) $17\frac{1}{2}°$, $17\frac{1}{2}°$

Use the figure below to answer questions 3 and 4.

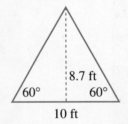

3. What is the perimeter of the triangle?

   (1) 18.7 ft   **(2)** 30 ft

   (3) 43.5 ft   **(4)** 87 ft

   (5) Insufficient information is given to solve this problem.

4. Find the area of the triangle.

   (1) 9.35 ft²   **(2)** 18.7 ft²

   (3) 43.5 ft²   **(4)** 87 ft²

   (5) Insufficient information is given to solve this problem.

5. Triangles *ABC* and *DEF* are similar. If *AB* is 4 in., *BC* is 6 in., and *EF* is 12 in., how long is *DE*?

   (1) 8 in.

   (2) 12 in.

   (3) 24 in.

   (4) 36 in.

   (5) 48 in.

6. If two angles of triangle *ABC* are equal to two angles of triangle *DEF*, _____.

   (1) $\triangle ABC \sim \triangle DEF$

   (2) $\triangle ABC \cong \triangle DEF$

   (3) the two triangles are the same

   (4) the two triangles may be similar

   (5) Insufficient information is given to solve this problem.

7. Find the area of a triangle with a 1-foot base and a 10-inch height.

   (1) 10 in. **(2)** 60 sq in.

   (3) 6 sq ft **(4)** 120 sq in.

   (5) 484 sq in.

8. Find the distance (*x*) across the lake.

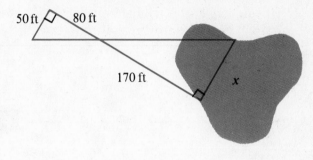

   (1) 40 ft **(2)** 23.5 ft

   (3) 28.2 ft **(4)** 27 ft

   (5) 106.25 ft

9. Find *x*.

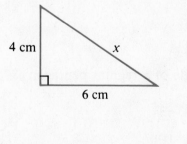

   (1) 7.21 cm **(2)** 10 cm

   (3) 12 cm **(4)** 20 cm

   (5) 52 cm

10. Find *y*.

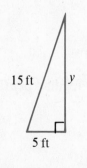

   (1) 10 ft **(2)** 14.14 ft

   (3) 15.8 ft **(4)** 20 ft

   (5) 200 ft

Pretest

Circle the best answer for each question.

1. What is a closed plane geometric figure having equal sides and equal angles called?

    (1) an equilateral triangle

    (2) an interior angle

    (3) a quadrilateral

    (4) a polygon

    (5) a regular polygon

2. What is a six-sided polygon called?

    (1) a hexagon

    (2) an octagon

    (3) a pentagon

    (4) a heptagon

    (5) a regular polygon

3. An interior angle of a regular pentagon measures _____ .

    (1) 36°   **(2)** 72°

    (3) 90°   **(4)** 108°

    (5) 900°

4. An exterior angle of a regular pentagon measures _____ .

    (1) 36°   **(2)** 72°

    (3) 90°   **(4)** 108°

    (5) 900°

5. The perimeter of a polygon is _____.

   (1) the part of the plane enclosed by the figure

   (2) the product of its length and its width

   (3) the distance around the figure

   (4) the sum of its interior angles

   (5) measured in square units

6. What is the perimeter of a regular octagon having sides measuring 5 cm?

   (1) 25 cm

   (2) 30 cm

   (3) 40 cm

   (4) 40 cm^2

   (5) Cannot be determined from the information given.

7. The perimeter of a hexagon is 8 in. What is the measure of one side?

   (1) Cannot be determined.

   (2) $1\frac{1}{3}$ in.

   (3) 1 in.

   (4) 48 in.

   (5) 64 in.

8. The area of a regular polygon is the product of _____.

   (1) one-half the base and the height

   (2) the length and the width

   (3) two times the length plus two times the width

   (4) the perimeter and the number of sides

   (5) one-half the perimeter and the apothem

9. What is the area of a regular pentagon having a side measuring 16 cm and an apothem measuring 11 cm?

   (1) 176 cm^2 (2) 88 cm^2

   (3) 440 cm^2 (4) 880 cm^2

   (5) 27 cm^2

10. A lot is in the shape of a regular hexagon with sides measuring 550 m. If the lot is divided into two equal parts, what is the perimeter of each part?

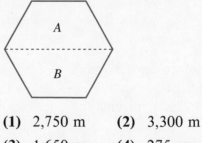

   (1) 2,750 m   (2) 3,300 m

   (3) 1,650 m   (4) 275 m

   (5) 5,500 m

LESSON 20

Angles and Sides of Polygons

Polygons are closed plane geometric figures. Some common polygons are shown in the chart below.

Name	**Number of Sides**	**Figure**
quadrilateral	four	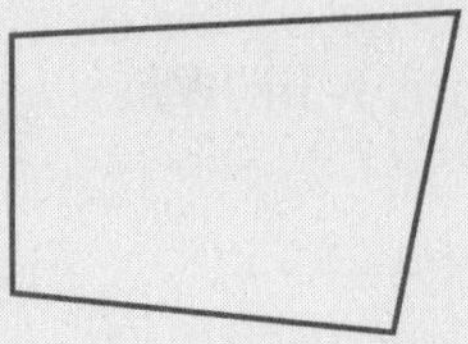
pentagon	five	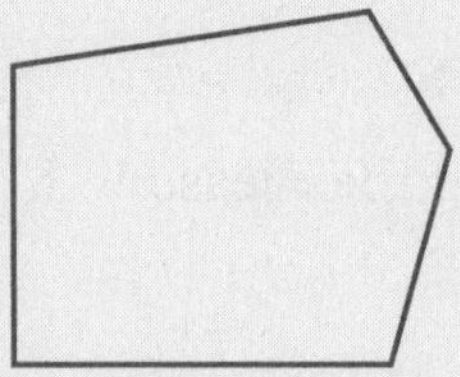
hexagon	six	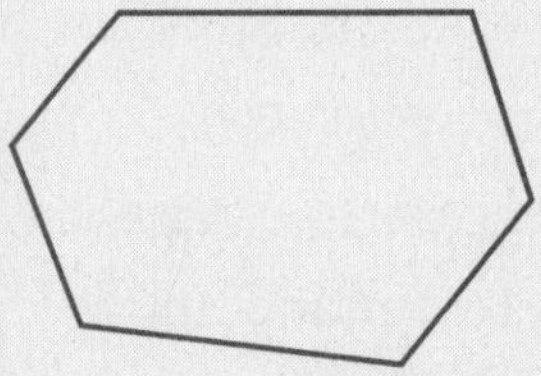
heptagon	seven	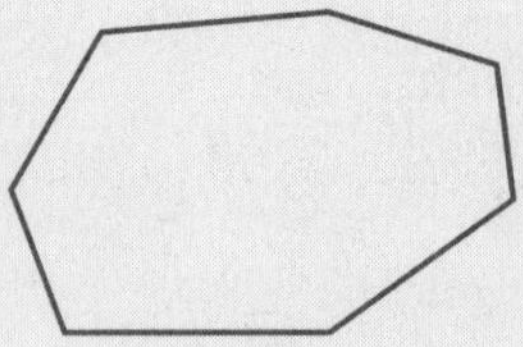
octagon	eight	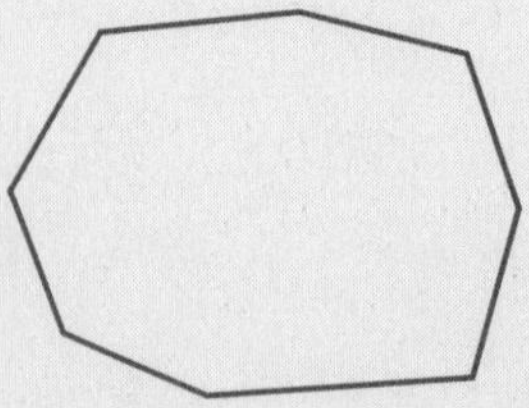

A **regular polygon** is a polygon that has equal sides and equal angles. You can calculate the measures of the inside angles of a regular polygon if you know how many sides the figure has.

To calculate the measures of the interior angles of a regular polygon, follow these steps:

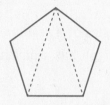

Step 1   Divide the figure into triangles.

Step 2   Count the number of triangles formed. The number of triangles formed in a polygon is always 2 less than the number of sides.

Step 3   Multiply this number by 180. Since the pentagon is divided into three triangles, the sum of the interior angles of a pentagon is 3(180)°, or 540°. If the pentagon is regular, then each of the angles has a measure of 540° ÷ 5, or 108°.

Therefore, the formula for finding the measure of the angles of a regular polygon is:

$$\frac{(n - 2)\,180}{n}$$

where n stands for the number of sides of the figure.

An exterior angle of a regular polygon is the supplement of an interior angle. The exterior angles are formed by extending the sides of the polygon.

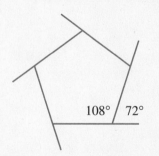

As noted before, each angle of a regular pentagon measures 108°. Since an exterior angle is the supplement of an interior angle, its measure is 180° − 108°, or 72°.

The sum of the exterior angles is 5 (72°), or 360°.

A. Find the sum of the interior angles of a regular hexagon. Then find the measure of each angle.

A hexagon can be divided into four triangles. $\quad (n - 2)$, or $(6 - 2) = 4$

$$4(180°) = 720°$$

The sum of the interior angles of a regular hexagon is 720°. Each of the angles of a regular hexagon has a measure of $\left(\frac{720°}{6}\right)$, or 120°.

B. Find the measures of the interior angles and exterior angles of a regular quadrilateral.

If we divide the quadrilateral into triangles, we get two triangles.

The measure of the interior angles is

$$\frac{(n - 2)\,180}{n}$$

where n stands for 4. So, each angle measures

$$\frac{(4 - 2)\,180}{4} = 90°$$

The measure of the exterior angles is $180° - 90°$, or 90°.

C. A regular polygon has an interior angle of 120°. How many sides does the polygon have?

$$\frac{(n - 2)\,180}{n} = 120$$
$$180n - 360 = 120n$$
$$60n = 360$$
$$n = 6$$

There are six sides.

For each regular polygon listed in problems 1–6, find (1) the measure of an interior angle, (2) the sum of the angle measures, (3) the measure of an exterior angle, and (4) the sum of the exterior angles.

1. equilateral triangle
   (1) _____   (2) _____
   (3) _____   (4) _____

2. quadrilateral
   (1) _____   (2) _____
   (3) _____   (4) _____

3. pentagon
   (1) _____   (2) _____
   (3) _____   (4) _____

4. hexagon
   (1) _____   (2) _____
   (3) _____   (4) _____

5. heptagon
   (1) _____   (2) _____
   (3) _____   (4) _____

6. octagon
   (1) _____   (2) _____
   (3) _____   (4) _____

7. Examine the results of questions 1–6. (1) What can you conclude about the size of the interior angles of a polygon as the number of sides increases? (2) What can you conclude about the sum of the exterior angles of a regular polygon?

   (1) _____

   _____

   (2) _____

   _____

8. The measure of an interior angle of a regular polygon is 156°. How many sides does the polygon have?

   _____

Geometric Designs

Floor tiles are cut into squares to make them easier to install. Other polygons are used in the design of many items found in the house, such as floors and countertops.

A design that looks like a honeycomb is actually a series of connected hexagons.

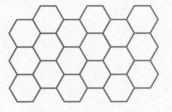

Name the type of polygon found in each floor design shown.

1.

2.

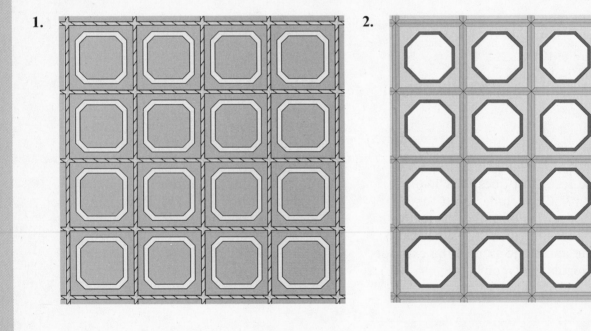

_____   _____

Perimeters of Polygons

The perimeter of a geometric figure is the distance around the figure. To find the perimeter of a figure, add all its sides together.

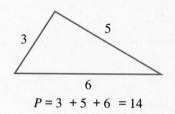

$$P = 3 + 5 + 6 = 14$$

--- **Examples** ---

A. Find the perimeter of a regular hexagon with sides measuring 5 inches.

A hexagon has 6 sides. Therefore, the formula is

$$P = 5 + 5 + 5 + 5 + 5 + 5 = 5(6)$$

The perimeter is 30 inches.

B. The perimeter of a regular hexagon is 45 cm. What is the measure of each side?

A hexagon has 6 sides.

$$45 \div 6 = 7.5$$

Each side has a measure of 7.5 cm.

--- **Practice** ---

Find the perimeter of each polygon.

1.

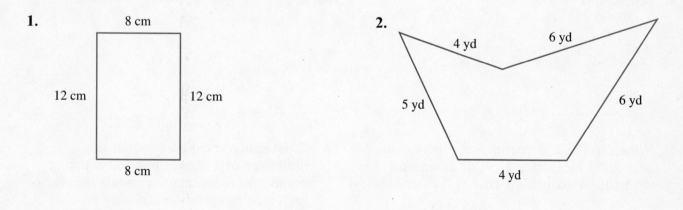

8 cm

12 cm   12 cm

8 cm

2.

4 yd   6 yd

5 yd   6 yd

4 yd

_____

Find the perimeter of each regular polygon. The variable s stands for the length of one side.

3. pentagon, s = 3.6 m

4. quadrilateral, s = 2 ft 6 in

_____

_____

5. octagon, s = 10 in.

6. triangle, s = 2.7 cm

_____

_____

Find the length of one side of each regular polygon. The variable P stands for perimeter.

7. pentagon, P = 325 ft

8. quadrilateral, P = 50.4 cm

_____

_____

9. pentagon, P = 18 m

10. octagon, P = 12m 8 cm

_____

_____

11. Which has a larger perimeter: a pentagon with sides measuring 1 cm or a hexagon with sides measuring 1 cm?

12. What can you conclude about the perimeter of a regular polygon if the number of sides increases while the size of each side remains the same?

_____

_____

Areas of Regular Polygons

The area of a polygon is the amount of surface it covers.

You can divide a regular polygon into triangles by drawing line segments from the center of the polygon to each vertex. To find the area of the polygon, you first find the areas of the triangles created. Then you add the areas together.

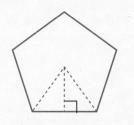

Since the sides of a regular polygon are equal and the line segments from the center to each vertex are equal, these triangles are **congruent.** The height of each triangle is called the **apothem.**

The sum of the areas of the triangles is the area of the regular polygon. This is represented by the following equation:

$$A = n\left(\tfrac{1}{2}\right)(bh)$$

where n = number of sides of the polygon, b = length of each side, and h = the apothem.

Example

Find the area of a pentagon with sides measuring 4 cm and the height of the triangle measuring 2.75 cm.

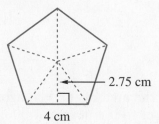

2.75 cm

4 cm

$$A = n\left(\tfrac{1}{2}\right)bh$$
$$A = 5\left(\tfrac{1}{2}\right)4(2.75)$$
$$A = 27.5$$

The area of the pentagon is 27.5 cm².

Find the area of each regular polygon described below.

1. hexagon, $b = 1$, apothem = 0.87 in.

2. heptagon, $b = 1$, apothem = 1.04m

_____

_____

3. octagon, $b = 1$, apothem = 1.21 m

4. 10-sided figure, $b = 1$, apothem = 1.54m

_____

_____

Problem Solving

Use the following information to solve problems 5 and 6:

The Pentagon in Arlington, Virginia, is where the Department of Defense is located. Each side of the building is 921 feet long. The apothem of the building is 633.8 feet.

5. Calculate the perimeter of the Pentagon.

6. Calculate the area of the Pentagon. Round your answer to the nearest 1,000 square feet.

_____

_____

Parts and Composites of Polygons

Sometimes you need to find the perimeter or area of only part of a polygon.

Example

Find the area of the shaded region of the regular pentagon shown below.

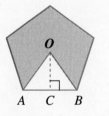

$OC = 24$ cm
$AB = 36$ cm

Note that four regions of the pentagon are shaded. To find the area of the shaded region, first find the area of one of the triangles formed. Then multiply the answer by 4.

Find the area of one triangle.

$$A = \tfrac{1}{2}bh$$
$$A = \left(\tfrac{1}{2}\right)(36)(24)$$
$$A = 432 \text{ cm}^2$$

Then multiply the area by 4 since four regions are shaded.

$$A = 4(432)$$

The area of the shaded region is 1,728 cm².

Practice

Solve each of the following.

1. Find the perimeter of the figure below. The figure is one half of a regular hexagon.

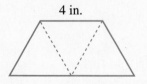

4 in.

2. Find the area of the shaded regions of the figure below. The figure is a regular hexagon.

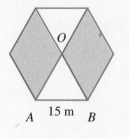

15 m

_____

_____

93

Problem Solving: Answering the Question

The steps you have learned to help solve word problems can be used with word problems that deal with geometric figures. Use the following steps:

Step 1 Read the problem and identify the key words. Underline the key words. These words will generally relate to some mathematical reasoning.

Step 2 Make a plan to solve the problem. Ask yourself, Should I add, subtract, multiply, divide, round, or compare? You may have to do more than one of these operations for the same problem.

Step 3 Find the solution. Use your math knowledge to find your answer.

Step 4 Check your answer. Ask yourself, Is the answer reasonable? Did you find what you were asked for?

When you solve problems, make sure you answer the question being asked. Sometimes the question being asked needs two answers. The equation may only give you one of those answers.

Example

A garden shaped like a regular hexagon needs a fence around it. Each side of the garden measures 6 feet. If the materials for the fencing cost $15 per foot, how much would the materials cost for the fence?

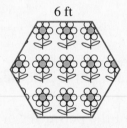

6 ft

Step 1 Read the problem and identify the key words. The key words are **around the garden** and **how much.**

Step 2 Make a plan to solve the problem. Find the perimeter of the garden. Multiply the perimeter by cost of materials.

Step 3 Find the solution.

$P = 6 + 6 + 6 + 6 + 6 + 6$
$P = 36$ feet

Finding the perimeter is not enough. You need to find the cost of the materials.

Since the cost of the materials is $15 per foot, you need to multiply the perimeter by 15.

Cost of the materials = $15(36)

The materials for the fence cost $540.

Step 4 Check the answer.

You needed to find the cost of the materials. Is the answer reasonable?

$$\$540 \div 36 = \$15$$

The numbers check.

Practice

Solve each of the following problems.

1. George wants to lay new tile in his bedroom. Each tile covers 1 square foot. His room is the shape of a regular pentagon with each side measuring 6 feet. The height of the apothem is 4.1 feet. How many tiles should he buy?

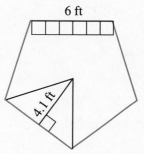

_____

2. Vanessa is going to put a patio in her back-yard. The figure shows the dimensions of the patio. Vanessa will lay bricks end-to-end all around the outside edges of the patio excluding the side that runs against the house. The bricks she uses are 10 inches long. How many bricks does she need to finish the patio?

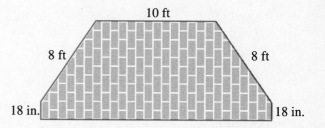

_____

Posttest

Circle the best answer for each question.

1. A regular polygon is a closed plane geometric figure with _____ .
   - (1) interior angles
   - (2) two or more sides
   - (3) equal sides
   - (4) equal angles
   - (5) both equal sides and equal angles

2. What is an eight-sided polygon called?
   - (1) a hexagon
   - (2) an octagon
   - (3) a pentagon
   - (4) a heptagon
   - (5) a regular polygon

3. The interior angles for a regular pentagon measure _____ .
   - (1) 108°   (2) 60°
   - (3) 90°   (4) 72°
   - (5) 360°

4. An interior angle of a regular polygon measures 160°. How many sides does the polygon have?
   - (1) 36   (2) 18
   - (3) 12   (4) 9
   - (5) 8

5. The sum of the exterior angles of a regular polygon is _____ .
   - (1) 360°
   - (2) 720°
   - (3) equal to the sum of the interior angles
   - (4) the supplement of one interior angle
   - (5) two times the sum of the interior angles

6. The distance around a polygon is _____ .
   - (1) the part of the plane enclosed by the figure
   - (2) the product of its length and its width
   - (3) the perimeter
   - (4) the sum of its interior angles
   - (5) measured in square units

7. What is the perimeter of a polygon with sides measuring 5 cm, 6 cm, 4 cm, and 3 cm?

   (1) 360 cm   **(2)** 180 cm

   (3) 18 cm   **(4)** 18 cm^2

   (5) Cannot be determined from the information given.

8. The perimeter of an octagon is 8 in. What is the measurement of one side?

   (1) Cannot be determined.   **(2)** 1 in.

   (3) $\frac{1}{2}$ in.   **(4)** 48 in.

   (5) 64 in.

9. What is the area of a regular pentagon having a side measuring 11 cm and an apothem measuring 18 cm?

   (1) 19.8 cm^2 **(2)** 99 cm^2

   (3) 198 cm^2 **(4)** 495 cm^2

   (5) 990 cm^2

10. A regular hexagon has sides measuring 10 cm. If the hexagon is divided into two equal parts, what is the area of each part?

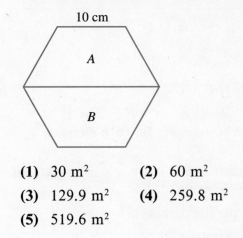

10 cm

A

B

   (1) 30 m^2 **(2)** 60 m^2

   (3) 129.9 m^2 **(4)** 259.8 m^2

   (5) 519.6 m^2

UNIT

5

Quadrilaterals

Circle the best answer for each question.

1. A quadrilateral has one pair of parallel sides. This quadrilateral is a _____ .

   (1) parallelogram
   (2) rectangle
   (3) square
   (4) rhombus
   (5) trapezoid

2. A quadrilateral has one pair of parallel sides. The sides that are not parallel are equal. What is this quadrilateral called?

   (1) a regular polygon
   (2) a rectangle
   (3) an isosceles trapezoid
   (4) a rhombus
   (5) a square

3. A quadrilateral has two pairs of parallel sides. This quadrilateral is a _____ .

   (1) parallelogram
   (2) rectangle
   (3) square
   (4) rhombus
   (5) trapezoid

4. Opposite angles of a parallelogram are _____ .

   (1) 90°
   (2) 45°
   (3) equal
   (4) complementary
   (5) supplementary

5. If one side of a rhombus is 9 cm, _____ .

   (1) the other sides each measure 9 cm

   (2) only the opposite side measures 9 cm

   (3) the other sides cannot be determined

   (4) the rhombus is a square

   (5) the other sides must be greater than 9 cm

6. One side of a parallelogram measures 19 yd. Another side measures 13 yd. What is the perimeter of the parallelogram in feet?

   (1) Cannot be determined.

   (2) 21.3 feet

   (3) 32 feet

   (4) 64 feet

   (5) 192 feet

7. One side of a parallelogram measures 19 yd. Another side measures 13 yd. What is the area of the parallelogram?

   (1) Cannot be determined.

   (2) 247 yd^2

   (3) 247 yd

   (4) 123.5 yd^2

   (5) 64 yd

8. The area of a rectangular plot of land is 60 ft.^{2} The width is 4 feet. What is the length in feet?

   (1) Cannot be determined.

   (2) 7.4 feet

   (3) 15 feet

   (4) 26 feet

   (5) 56 feet

9. What is the area of this trapezoid?

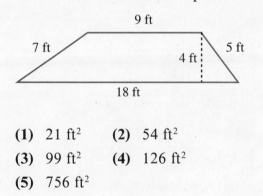

   (1) 21 ft^2 (2) 54 ft^2

   (3) 99 ft^2 (4) 126 ft^2

   (5) 756 ft^2

10. What is the surface area of this wall?

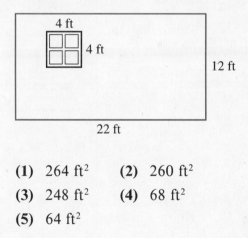

   (1) 264 ft^2 (2) 260 ft^2

   (3) 248 ft^2 (4) 68 ft^2

   (5) 64 ft^2

Angles and Sides of Quadrilaterals

As you have already learned, a quadrilateral is a four-sided figure. An interior angle of a regular quadrilateral measures 90°. The sum of the interior angles of a regular quadrilateral is 360°.

You can divide any quadrilateral into two triangles by connecting two vertices that are not next to each other. This line segment is called a **diagonal.**

Since the sum of the angles of a triangle is always 180°, the sum of the angles of any quadrilateral will be 360°. In a quadrilateral that is not regular, the angles are not equal.

As shown on the chart below, some quadrilaterals have special names.

Name	Figure	Characteristics
Trapezoid		One pair of parallel sides called bases. $\overline{AB} \parallel \overline{CD}$
Parallelogram		Two pairs of parallel sides. $\overline{AB} \parallel \overline{CD}, \overline{BC} \parallel \overline{AD}$ Parallel sides are equal. $\overline{AB} = \overline{CD}, \overline{BC} = \overline{AD}$ Opposite angles are equal. $\angle A = \angle C, \angle B = \angle D$
Rhombus	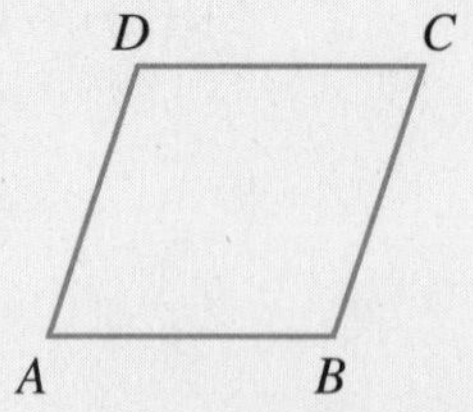	A parallelogram with four equal sides. $\overline{AB} \parallel \overline{CD}, \overline{BC} \parallel \overline{AD}$ $\overline{AB} = \overline{CD} = \overline{BC} = \overline{AD}$ $\angle A = \angle C, \angle B = \angle D$

Name	Figure	Characteristics
Rectangle	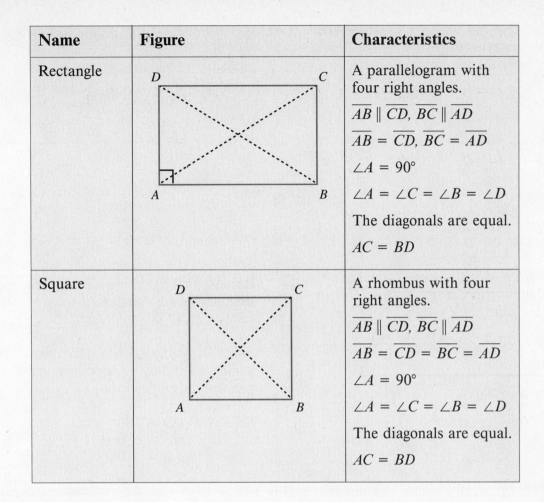	A parallelogram with four right angles. $\overline{AB} \parallel \overline{CD}$, $\overline{BC} \parallel \overline{AD}$ $\overline{AB} = \overline{CD}$, $\overline{BC} = \overline{AD}$ $\angle A = 90°$ $\angle A = \angle C = \angle B = \angle D$ The diagonals are equal. $AC = BD$
Square		A rhombus with four right angles. $\overline{AB} \parallel \overline{CD}$, $\overline{BC} \parallel \overline{AD}$ $\overline{AB} = \overline{CD} = \overline{BC} = \overline{AD}$ $\angle A = 90°$ $\angle A = \angle C = \angle B = \angle D$ The diagonals are equal. $AC = BD$

If you know the dimensions of a rectangle or a square, you can use the Pythagorean Theorem to find the length of the diagonal.

Example

A. A square has sides measuring 2 inches. Find the length of the diagonal.

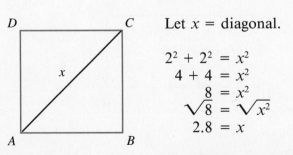

Let x = diagonal.

$$2^2 + 2^2 = x^2$$
$$4 + 4 = x^2$$
$$8 = x^2$$
$$\sqrt{8} = \sqrt{x^2}$$
$$2.8 = x$$

The diagonal is 2.8 inches long.

A trapezoid that has base angles which are equal is an **isosceles trapezoid.** The legs of an isosceles trapezoid are equal also.

Example

B. The figure below is an isosceles trapezoid. Find the measure of each angle.

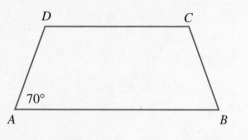

Because this figure is an isosceles trapezoid, the base angles are equal. Since $\angle A = 70°$, you know $\angle B = 70°$.

Since $\overline{DC} \parallel \overline{AB}$, $\overline{AD}$ is a transversal. Interior angles created by a transversal are supplements.

So, $\angle A + \angle D = 180$
$70 + \angle D = 180$
$\angle D = 110$

The sum of the interior angles of a quadrilateral is 360°.
Therefore, $\angle A + \angle B + \angle C + \angle D = 360°$.
By substitution, $70° + 70° + \angle C + 110° = 360°$.
So, $\angle C = 110$.

Practice

For each figure below, find the measurements of the missing angles.

1. Figure *ABCD* is a parallelogram.

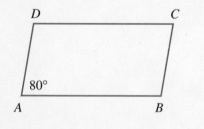

2. Figure *RSTU* is a rectangle.

3. Figure *EFGH* is a rhombus.

_____

4. Figure *WXYZ* is an isosceles trapezoid.

_____

5. Find the length of the diagonal *AC* of rectangle *ABCD*.

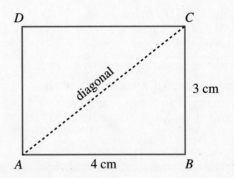

_____

Perimeters of Quadrilaterals

The perimeter of a figure is the sum of its sides.

Examples

A. Find the perimeter of trapezoid *DEFG* with sides measuring 1.5 cm, 2 cm, 2.5 cm, and 4 cm.

The perimeter is 10 cm.

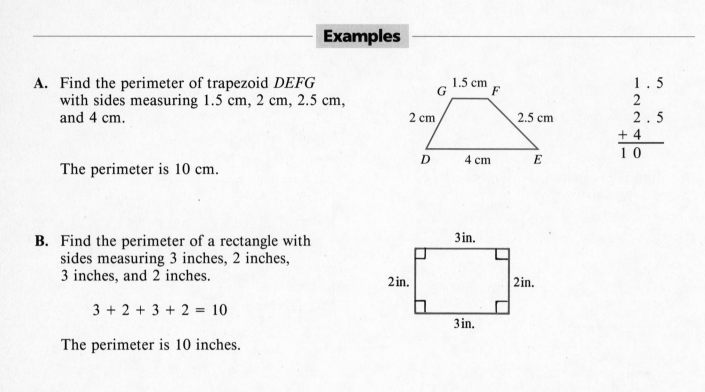

$$\begin{array}{r} 1.5 \\ 2 \\ 2.5 \\ +\ 4 \\ \hline 10 \end{array}$$

B. Find the perimeter of a rectangle with sides measuring 3 inches, 2 inches, 3 inches, and 2 inches.

$$3 + 2 + 3 + 2 = 10$$

The perimeter is 10 inches.

A formula that also can be used for the perimeter of a rectangle is

$$P = 2l + 2w$$

where *P* stands for perimeter, *l* stands for the rectangle's length, and *w* stands for the rectangle's width.

$$P = 2(3) + 2(2)$$
$$P = 6 + 4$$
$$P = 10$$

Remember, a square has the same measure for every side.

MATH HINT

The side of a rectangle with the longer measure is the length and the side with the shorter measure is the width.

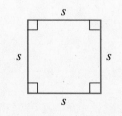

The formula for the perimeter of a square is $P = 4s$, where s stands for the measure of the sides.

Example

C. Find the perimeter of quadrilateral *ABCD*. *ABCD* is a parallelogram. *AB* measures 5 cm, and *BC* measures 3.5 cm.

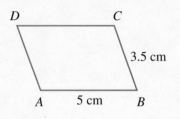

Remember, opposite sides of a parallelogram are equal. $AB = 5$ cm; so, $DC = 5$ cm. $BC = 3.5$ cm; so, $AD = 3.5$ cm.

$P = 5 + 3.5 + 5 + 3.5$
$P = 17$

The perimeter of the figure is 17 cm.

Using the characteristics of a figure, you can find missing measures.

D. The perimeter of a rhombus is 105 m. Find the measure of each side.

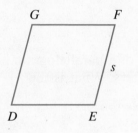

Recall that all the sides of a rhombus are equal. The perimeter can be found using the formula $P = 4s$, where s is the measure of each side. So,

$105 = 4s$
$26.25 = s$

Each side measures 26.25 m.

Practice

Answer each of the following questions.

1. Measure the cover of this book.

   (1) What is its length?   _____

   (2) What is its width?   _____

   (3) What is its perimeter?   _____

   (4) Name the quadrilateral.   _____

2. Measure a door.

   (1) What is its length (height)?   _____

   (2) What is its width?   _____

   (3) What is its perimeter?   _____

   (4) Name the quadrilateral.   _____

Find the perimeters of these quadrilaterals.

3.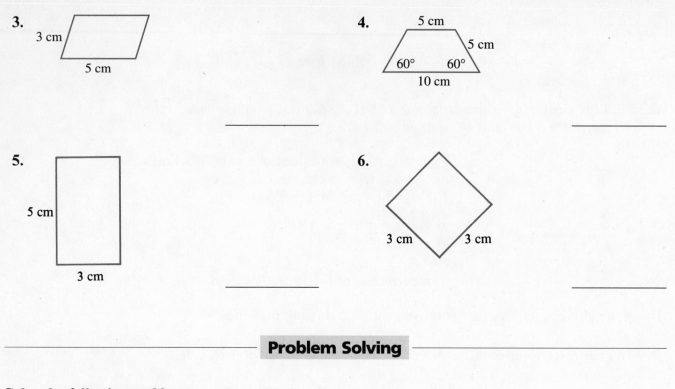

3 cm

5 cm

_____

4. 5 cm

5 cm

60° 60°

10 cm

_____

5.

5 cm

3 cm

_____

6.

3 cm 3 cm

_____

Problem Solving

Solve the following problems.

7. What is the perimeter of a rectangle if its length is 6 feet and its width is 4 feet?

_____

8. The adjacent sides of a parallelogram are 3 meters and 9 meters. What is the perimeter?

_____

9. How many meters of fencing are needed to enclose a rectangular field 150 meters by 80 meters?

_____

10. How many feet of baseboard are needed for a 20-foot by 16-foot room? (Subtract a total of 5 feet for doorways.)

_____

11. The perimeter of a square ceramic tile is 32 inches. What is the measure of each side of the tile?

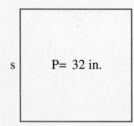

s   P= 32 in.

_____

12. A rectangular pasture required 5,000 m of fencing. The shorter side measures 800 m. What is the measure of the longer side?

800 m

_____

Areas of Quadrilaterals

The area of a polygon is the surface it covers. Area is written in square units.

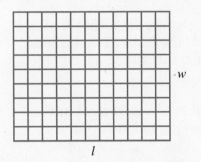

Look at the figure.
Count the number of square units.
There are 99 square units.

The area of a rectangle can be found by using the formula

$$A = lw$$

where A stands for area, l stands for length, and w stands for width.

In the rectangle shown, the length is 11 units; the width is 9 units. So,

$$A = lw$$
$$A = 11(9)$$
$$A = 99 \text{ square units}$$

Example

A. Find the area of a rectangle with a length of 1 foot and a width of 6 inches.

Change 1 foot to 12 inches.

$$A = lw$$
$$A = 12(6)$$
$$A = 72$$

The area of the rectangle is 72 square inches.

You could also have changed the 6 inches to 0.5 feet, or to $\frac{1}{2}$ ft.

$$A = lw$$
$$A = 1(0.5)$$
$$A = 0.5$$

The area of the rectangle is 0.5 square feet.

> **MATH HINT**
>
> **A**ll units must be changed to the same unit before you use any formulas.

> **MATH HINT**
>
> **1** square foot is the same as 144 square inches. Therefore, 72 square inches are one-half of a square foot.

The chart below shows formulas for finding the areas of other quadrilaterals.

Figure	Formula	Description of Formula
Square	$A = s^2$	The variable s is the measure of each side.
Parallelogram	$A = bh$	The variable b is the measure of the base of the figure. The variable h is the measure of the height of the figure.
Trapezoid	$A = \frac{1}{2}(b_1 + b_2)h$	The variable h is the height of the figure, b_1 is the measure of the top base of the figure, and b_2 is the measure of the bottom base of the figure.

Examples

B. Armand is covering a rectangular floor with parquet tiles. The floor is 10 feet long and 15 feet wide. Each tile is a square with 6-inch sides. How many tiles does Armand need for the floor?

First, find the areas of both the tiles and the floor.

The formula for finding the area of the floor is:

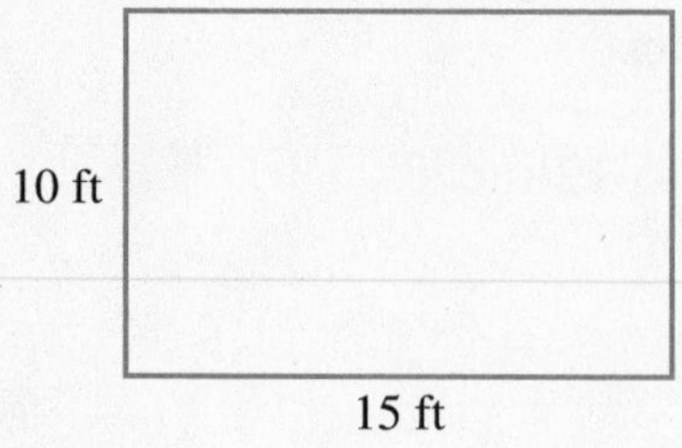

$A = lw$
$A = 10(15)$
$A = 150$

10 ft

15 ft

The area of the floor is 150 square feet.

The formula for finding the area of each tile is $A = lw$.

$A = 6(6)$
$A = 36$

The area of each tile is 36 square inches.

Next, divide the area of the tiles into the area of the floor. Use the same units of measure for both.

36 sq. in. = 0.25 sq. ft.
150 sq. ft ÷ 0.25 sq. ft = 600

Armand needs at least 600 tiles.

C. The area of a rectangle is 150 square feet. Its length measures 25 feet. Find the measurement of its width. Recall that the formula for finding the area of a rectangle is:

$A = lw$
$150 = 25\ w$
$6 = w$

The rectangle measures 25 feet long by 6 feet wide.

Practice

Find the area of each figure.

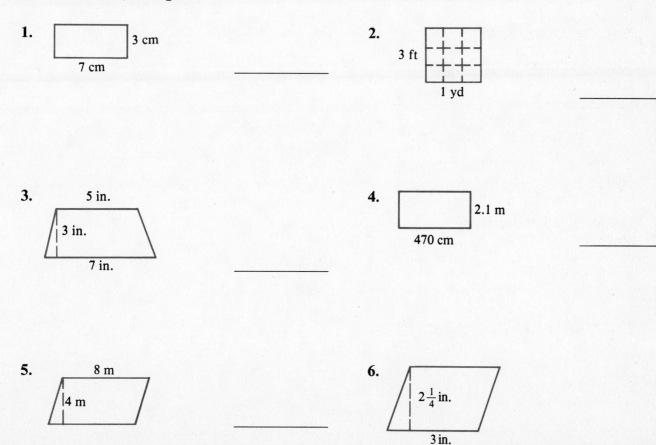

1.

3 cm
7 cm

_____

2.

3 ft
1 yd

_____

3.

5 in.
3 in.
7 in.

_____

4.

2.1 m
470 cm

_____

5.

8 m
4 m

_____

6.

$2\frac{1}{4}$ in.
3 in.

_____

Solve the following problems.

7. How many square feet of carpet are needed for a hallway 32 feet long and 3 feet wide?

8. How many square feet will a sign painter have painted if a billboard is 12 feet by 8 feet?

9. Allowing one hundred square feet for each tree, how many trees can be planted in an orchard 325 feet by 200 feet?

10. Allowing 30 square feet per child, about how many children can be permitted to play in a play area 80 feet by 100 feet?

Parts and Composites of Quadrilaterals

Some figures are combinations of several polygons. Sometimes you will need to find the areas of several polygons in order to find the area of a surface.

Examples

A. How much paint is needed to paint the wall shown below, if one quart of paint covers 100 square feet?

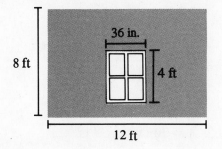

Step 1 Find the area of the wall.

$$A = lw$$
$$A = 12(8)$$
$$A = 96 \text{ square feet}$$

Step 2 Find the area of the window.

$$A = 4(3)$$
$$A = 12 \text{ square feet}$$

Step 3 Subtract the area of the window from the area of the wall.

$$96 - 12 = 84$$

The surface to be painted measures 84 square feet.

Step 4 Divide the total area of the wall by the area covered by one quart of paint. You find $84 \div 100 = 0.84$. You need only one quart of paint to cover the wall.

B. How much carpet does an L-shaped living/dining room area need?

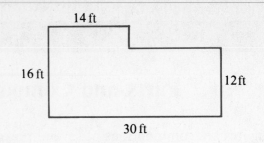

Step 1 Divide the shape of the room into smaller figures for which you know the formula for finding the area. The room can be divided several different ways. One way is shown below.

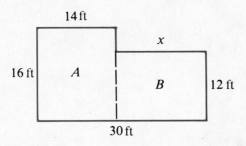

Step 2 Find the area of rectangle *A*. Use the formula *A = lw,* where *l* is 16 and *w* is 14.

Area of *A* = 16(14), or 224 square feet.

Step 3 Find the area of rectangle B. Use the formula, *A = lw,* where *l* is (30 − 14), or 16; and *w* is 12.

Area of B = 16(12), or 192 square feet.

Step 4 The total area of the room is 224 + 192, or 416 square feet. Most carpet is measured in square yards. To change square feet to square yards, divide 416 by 9. (Hint: 9 square feet = 1 square yard.) To carpet the room, you would need $46\frac{2}{9}$ square yards.

Solve each of the following problems.

1. How many square feet of Formica are needed to cover this countertop?

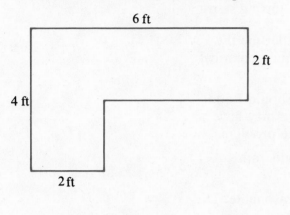

2. A room has the shape and dimensions shown below. How much baseboard is needed? (Subtract 9 feet for doors.)

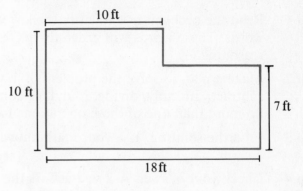

_____

3. How many feet of fencing are needed to fence this yard?

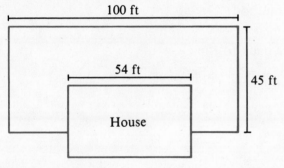

4. How many square feet of sod are needed to cover this yard?

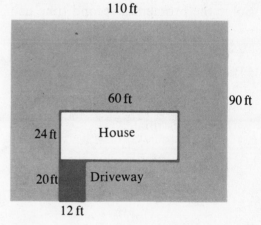

_____

5. How many rolls of unpatterned wallpaper are needed to paper this wall? Each roll of wallpaper covers 40 square feet.

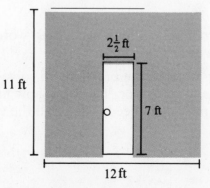

_____

Problem Solving: Check the Answer

The steps you have learned to solve word problems can be used with word problems that deal with geometric figures. Use the following steps:

Step 1 Read the problem and identify the key words. Underline the key words. These words will generally relate to some mathematical reasoning.

Step 2 Make a plan to solve the problem. Ask yourself, Should I add, subtract, multiply, divide, round, or compare? You may have to do more than one of these operations for the same problem.

Step 3 Find the solution. Use your math knowledge to find your answer.

Step 4 Check your answer. Ask yourself, Is the answer reasonable? Did you find what you were asked for?

All problems you solve need to be checked to make sure the answers are reasonable. To check your work you can do the following:

1. Solve the problem a second time using a different method.

2. Work the problem in reverse order to see if you get back to the original numbers.

3. Estimate high and low answers to see if the actual answer falls within the estimate.

Example

Weather stripping costs $1.29 for each 6-foot length and is sold only in the 6-foot strips. How much will it cost to weatherstrip a door 36 by 78 inches?

Step 1 Read the problem and identify the key words.
The key words are **how much.**

Step 2 Make a plan to solve the problem. To solve this problem, find the perimeter of the door. Then find how many strips of the weather stripping you need. Then find the cost of the weather stripping.

Step 3 Find the solution. A door is a rectangle, so you can use the
formula, $P = 2(l) + 2(w)$.

$$P = 2(78) + 2(36)$$
$$P = 156 + 72$$
$$P = 228$$

The perimeter of the door is 228 inches.

The weather stripping is sold in 6-foot lengths. You need to
change 228 inches to feet.

$$228 \div 12 = 19$$

The perimeter of the door is 19 feet.

To find the number of strips you need, divide 19 by 6.

$$19 \div 6 = 3.167$$

You must buy the strips in 6-foot lengths. You will need to buy
4 of the 6-foot strips. Each strip costs $1.29.

$$4(\$1.29) = \$5.16$$

It will cost $5.16 for materials for the weather stripping.

Step 4 Check your answer.

Change the dimensions of the door to feet. The perimeter of the
door is $2(3) + 2(6.5)$, or 19 feet. This checks with the previous
calculations.

$$6 \text{ ft} \times 3 \text{ strips} = 18 \text{ ft.}$$
$$6 \text{ ft} \times 4 \text{ strips} = 24 \text{ ft.}$$

Since the perimeter is between 18 and 24, you will need more
than 3 strips and less than 4 strips. You must buy 4 strips. This
checks also.

The answer is reasonable.

1. How much will it cost to put a double row of fringe around a bedspread which is a square with sides measuring 260 cm? (Fringe is not sewn across the top of the spread.) Fringe costs $.70 per meter.

   _____

2. How many times must Josh jog around a square city block $\frac{1}{4}$ mile on each side to jog 3 miles?

   _____

3. Sandra needs to fertilize her garden. Two pounds of fertilizer are enough for 100 square feet. Use the measurements below to find out how much fertilizer she needs.

4. How many acres does a rancher own if his property is in the shape of a trapezoid as shown? (One acre equals 45,560 square feet.)

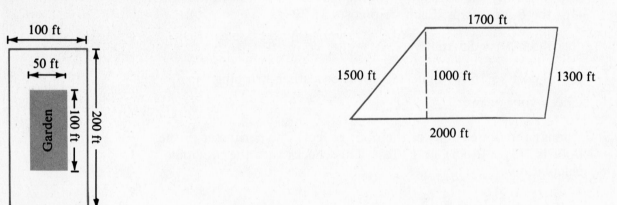

   _____   _____

5. How many times can you wrap a string 42 feet long around a box measuring $2\frac{1}{2}$ feet on each side?

   _____

Finding the Lowest Rental Cost

The cost of renting a building is found by multiplying the cost per square foot by the number of square feet in the building.

The Lake View Neighbors Association needs to rent office space. The committee found two possibilities: a store front and a small factory. The store front is 3,500 square feet and the rent for the year is $2.50 a square foot. The factory is 4,500 square feet and the rent is $1.75 a square foot.

For the store front, the annual rent would be
$2.50 \times 3,500 = \$8,750$
The monthly payment would be
$\$8,750 \div 12 = \$729.17$

Solve.

1. What is the total annual rent for the factory? _____

2. What is the monthly rent for the factory? _____

3. The heat for the store front would be less if the two windows could be covered in plastic and had curtains. Each window is 4 feet wide × 9 feet high. A roll of plastic is 50 inches wide and 10 feet long and costs $10.99. Material for curtains is 54 inches wide and costs $4.99 a yard. The curtains will need to be turned under 6 inches at the top and bottom. How much will it cost to cover the windows? _____

4. The factory is one large room. To make an office, the group will have to build a room 20 feet long, 15 feet wide, and 8 feet tall. Two of the factory walls can be used for the office, so only two new walls need to be built. Sheetrock costs $11.99 for each 4 feet × 8 feet panel. Supports for the walls are $3 each and are placed at each end and 18 inches apart. How much will it cost to build the office? _____

5. Which building will be cheaper to rent for the first year? Which will cost more to move into? _____

Circle the best answer for each question.

1. One pair of opposite sides of a quadrilateral are parallel. What is this quadrilateral called?

   (1) a parallelogram

   (2) a rectangle

   (3) a square

   (4) a rhombus

   (5) a trapezoid

2. All of the angles and sides of a quadrilateral are equal. What is this quadrilateral called?

   (1) a regular polygon

   (2) a rectangle

   (3) an isosceles trapezoid

   (4) a rhombus

   (5) a square

3. A quadrilateral has two pairs of equal sides. What is this quadrilateral called?

   (1) a parallelogram

   (2) a rectangle

   (3) a square

   (4) a rhombus

   (5) a trapezoid

4. Opposite angles of a rhombus are _____.

   (1) 90°

   (2) 45°

   (3) equal

   (4) complementary

   (5) supplementary

5. If one side of a parallelogram is 9 cm, _____.

   (1) the other sides each measure 9 cm

   (2) only the opposite side measures 9 cm

   (3) the other sides cannot be determined

   (4) the parallelogram is a square

   (5) the other sides must be greater than 9 cm

6. One side of a rectangle measures 57 ft. Another side measures 39 ft. What is the perimeter of the rectangle in yards?

   (1) Cannot be determined.

   (2) 576 yards

   (3) 192 yards

   (4) 96 yards

   (5) 64 yards

Solve.

7. One side of a parallelogram measures 57 feet. Another side measures 39 feet. The height of the parallelogram is 36 feet. What is the area of the parallelogram?

   (1) Cannot be determined.

   (2) 2,223 ft²

   (3) 2,052 ft²

   (4) 1,111.5 ft²

   (5) 192 ft²

8. The area of a square plot of land is 60 ft.² What is the length of each side in feet?

   (1) Cannot be determined.

   (2) 7.7 feet

   (3) 15 feet

   (4) 26 feet

   (5) 56 feet

9. What is the perimeter of this trapezoid?

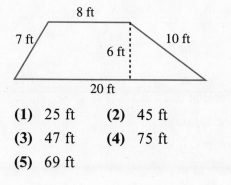

   (1) 25 ft   (2) 45 ft

   (3) 47 ft   (4) 75 ft

   (5) 69 ft

10. What is the surface area of this wall?

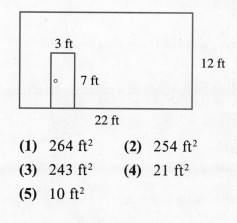

   (1) 264 ft²   (2) 254 ft²

   (3) 243 ft²   (4) 21 ft²

   (5) 10 ft²

Pretest

Circle the best answer for each question.

1. The radius of a circle is 9 inches. How would you find the diameter of the circle?

   (1) Multiply by 2.

   (2) Divide by 2.

   (3) Multiply by π.

   (4) Multiply by 0.5.

   (5) Divide by the circumference.

2. An angle whose vertex is at the center of a circle is called _____ .

   (1) an acute angle

   (2) an obtuse angle

   (3) a central angle

   (4) a right angle

   (5) a radius

3. The circumference of a circle is _____ .

   (1) the diameter

   (2) the radius

   (3) one-half the circle

   (4) the distance around the circle

   (5) a straight line

4. The area of a circle is 50.24 square meters. What is its diameter?

   (1) 4 meters

   (2) 8 meters

   (3) 16 meters

   (4) 2 meters

   (5) 10 meters

5. The diameter of a circle is 14 cm. What is its circumference?

    (1) 22 cm

    (2) 44 cm

    (3) 87.92 cm

    (4) 88 cm

    (5) 616 cm

6. A circle has a radius of 7 cm. What is its area?

    (1) 616 cm^2

    (2) 313 cm^2

    (3) 196 cm^2

    (4) 154 cm^2

    (5) 87.92 cm^2

Problem Solving

Solve the following problems.

7. How many feet of edging will be needed for a round table with a radius of $2\frac{1}{2}$ feet?

8. Two metal hoops are to be fitted around a barrel. Each hoop will have a diameter of 2 feet. Allow 6 inches for fastening the ends of each hoop. How many feet of metal are needed?

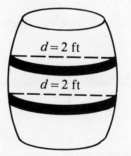

$d = 2$ ft

$d = 2$ ft

_____

_____

Parts and Angles of Circles

A **circle** is a curve that starts and ends at the same point and does not cross itself. All points on this curve are the same distance from the center.

Note that a circle does not include any interior points or the center point.

A circle is named by its center. The circle to the left is called circle *O*.

A **diameter** is a straight line passing through the center of a circle. The diameter connects two points on a circle.

A **radius** is a straight line connecting the center of a circle to a point on the circle. The radius of a circle is one-half the length of the diameter. The plural form of the word *radius* is *radii*.

A line segment from one point on a circle to another is called a **chord**.

The distance around the circle is its **circumference.** It is like the perimeter of a polygon.

An angle that has its vertex at the center of the circle is a **central angle.**

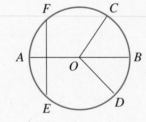

Practice

Refer to the circle to answer questions 1–6.

1. Name two radii. _____ _____

2. Name a diameter. _____

3. Name two chords. _____ _____

4. Name a central angle. _____

5. If the radius equals 5 inches, find the diameter. _____

6. If the diameter equals 30, find the radius. _____

Circumferences of Circles

The distance around a circle is its circumference. Recall that you can find the perimeter of a polygon by adding the lengths of its sides.

$(P = 2l + 2w)$. The perimeter of rectangle $ABCD$ is $2(10) + 2(4)$, or 28.

You cannot add the lengths of the sides of a circle because a circle has no sides.

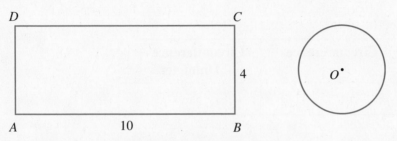

The formula for finding the circumference of a circle is $C = \pi d$ or $C = 2\pi r$, where d is the diameter and r is the radius of the circle.

The Greek letter π is pronounced "pi (long i)." This number is a ratio that stays constant. π is approximately 3.14, or $\frac{22}{7}$.

Some calculators have a key labeled π. You can press this key when you want to use the number π.

Examples

A. Find the circumference of a circle with a diameter of 21 inches.

$C = \pi d$
$C = \frac{22}{7}(21)$
$C = 66$

The circumference is 66 inches.

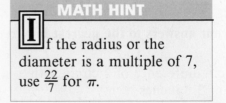

MATH HINT

If the radius or the diameter is a multiple of 7, use $\frac{22}{7}$ for π.

B. Find the circumference of a pulley with a radius of 7.5 cm.

$C = 2\pi r$
$C = 2(3.14)(7.5)$
$C = 47.1$

The circumference is about 47.1 cm.

C. The circumference of a circle is 48 cm. Find the radius.

$$C = 2\pi r$$
$$48 = 2(3.14)r$$
$$48 = 6.28r$$
$$\frac{48}{6.28} = r$$
$$7.64 = r$$

The radius is 7.64 cm.

Practice

Complete the following table.

	Radius	Diameter	Circumference	Circumference ÷ Diameter
1.		10		
2.	4			
3.			62.8	
4.			110	

5. Study the completed table. What is the ratio of the circumference to the diameter in each case?

_____

Problem Solving

Solve. Round your answers to the nearest hundredth.

6. Find the circumference of a circle with a diameter of 2.5 inches.

7. Find the circumference of a circle with a radius of 1.5 inches.

_____   _____

8. A wheel has a diameter of 3.5 feet. How many feet will it travel in one turn?

_____

9. A circular parking lot is 217 feet in diameter. How many feet of fencing will be needed to go around it?

_____

10. A bolt has a diameter of $\frac{3}{16}$ inch. Will it fit in a hole with a radius of 0.0465 inch?

_____

11. The radius of a bicycle tire is 13 inches. How far will the bike travel in two turns of the tire?

_____

LIFE SKILL

Circle Charts

Charts and graphs show information quickly and clearly. A common graph is a circle, or pie, graph. This type of graph compares parts of a whole to the whole.

This circle graph shows sources of income for a small company.

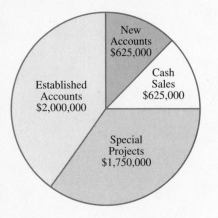

To draw a circle graph, you must first write a ratio that compares the parts of a whole to the whole. For example, if the total income is $5,000,000 and Special Projects is for $1,750,000, the ratio of

$$\frac{\text{Parts of the Whole}}{\text{Whole}} \text{ is } \frac{1,750,000}{5,000,000}, \text{ or } 35\%.$$

A complete circle has 360°. To find the number of degrees of the central angle for Special Projects, multiply 35% by 360°.

$$360(0.35) = 126$$

The measurement of the central angle is 126°.

Draw a circle graph showing the expenses of a small company. Use the following percents for spending:

Inventory	30%
Employee Benefits and Salaries	40%
Payment for Loans	10%
Operating Expenses	20%

The measurements of the central angles for the expense categories are as follows:

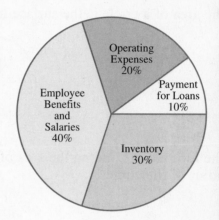

Inventory = 360(0.30)
            = 108
The measurement of the central angle is 108°.

Employee Benefits = 360(0.40)
                    = 144
The measurement of the central angle is 144°.

Payment for Loans = 360(0.10)
                    = 36
The measurement of the central angle is 36°.

Operating Expenses = 360(0.20)
                     = 72
The measurement of the central angle is 72°.

If you know the measurements of the central angles, you can use a protractor to draw a circle graph.

Use the circle graph above to answer each of the following questions.
Total expenses are $4,500,000.

1. How much was spent for Employee Benefits?

   _____

2. How much was spent on paying back loans?

   _____

Areas of Circles

The area of a circle is the surface inside the circle.

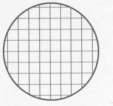

The formula for finding the area of a circle is $A = \pi r^2$, where r is the radius of the circle.

Examples

A. Find the area of a circle with a radius of 5 inches.

$$A = \pi r^2$$
$$A = 3.14(5)(5)$$
$$A = 78.5$$

The area of the circle is 78.5 square inches.

B. Find the area of a circle with a diameter of 5 inches.

$$A = \pi r^2$$
$$A = 3.14(2.5)(2.5)$$
$$A = 19.625$$

The area is about 19.625 square inches.

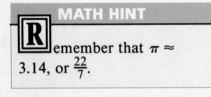

MATH HINT

Remember that $\pi \approx$ 3.14, or $\frac{22}{7}$.

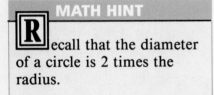

MATH HINT

Recall that the diameter of a circle is 2 times the radius.

C. The area of a circle is 50 cm². Find the diameter.

$$A = \pi r^2$$
$$50 = 3.14 r^2$$
$$\frac{50}{3.14} = r^2$$
$$\sqrt{15.92} = r$$
$$3.99 = r$$

Since the diameter is 2 times the radius, the diameter is about 7.98 cm.

───────────────────────── **Practice** ─────────────────────────

Complete the following table.

	Radius	Diameter	Area
1.	7 cm		
2.	1.5 cm		
3.		18 yd	
4.		10 ft	
5.		7 m	
6.	$5\frac{2}{3}$ in.		
7.		$8\frac{1}{6}$ mi	
8.		4.5 km	
9.			$\frac{11}{14}$ sq mi
10.			0.43 sq m

Solve.

11. A lawn sprinkler sprays water 15 feet in all directions. What is the area of lawn that can be covered by the sprinkler?

    _____

12. A high tower enables forest rangers to see for a distance of 45 km in all directions. How much area can be watched?

    _____

13. A 5-foot diameter round tabletop is to be covered with glass. How many square feet of glass are needed?

    _____

14. A television station can transmit over a radius of 60 miles. How large an area can receive its programs?

    _____

15. How much fertilizer is needed for a circular flower bed with a radius of 9 feet if one pound covers 100 square feet? Round to nearest tenth.

    _____

16. About how many people will a cake 10 inches in diameter serve if 6 square inches are allowed for each person?

    _____

Parts and Composites of Circles

Sometimes circumference and area problems involve part of a circle or several circles. To solve problems involving composite figures, divide the figure into shapes that you can use to find the areas.

Example

Find the area of this figure.

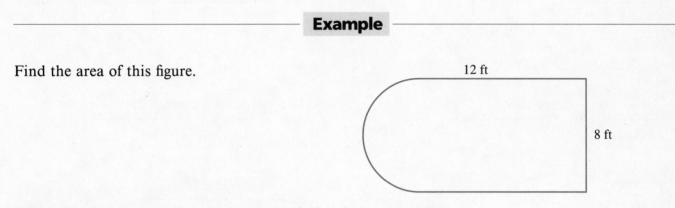

The figure divides into a rectangle and half of a circle, also called a semicircle.

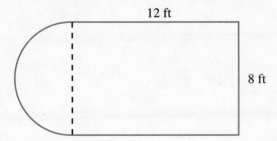

To find the area of the rectangle, use the formula, $A = lw$.

$$A = 12(8)$$
$$A = 96$$

The area of the rectangle is 96 sq ft.

To find the area of the semicircle, find the area of a complete circle that has a radius of 4 feet ($8 \div 2$). Use the formula $A = \pi r^2$. Divide that answer by 2 to find the area of the semicircle.

$$A = 3.14(4)(4)$$
$$A = 50.24$$

The area of the complete circle is 50.24 sq ft. The area of the semicircle is $50.24 \div 2$, or 25.12 sq ft.

To find the area of the total figure, add the areas of the rectangle and the semicircle together.

$$96 + 25.12 = 121.12$$

The area of the total figure is 121.12 sq ft.

Find the perimeter and area for each figure. Round answers to the hundredths place.

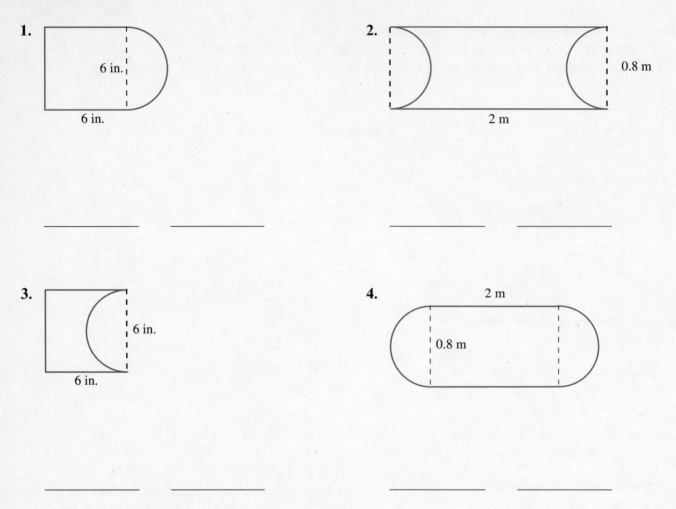

1.

6 in.

6 in.

_____ _____

2.

0.8 m

2 m

_____ _____

3.

6 in.

6 in.

_____ _____

4.

2 m

0.8 m

_____ _____

Problem Solving: Using Parts and Composites

The steps you have learned to solve word problems can be used with word problems that deal with geometric figures. Use the following steps:

Step 1   Read the problem and underline the key words. These words will generally relate to some mathematical reasoning.

Step 2   Make a plan to solve the problem.
Ask yourself, Should I add, subtract, multiply, divide, round, or compare? You may have to do more than one of these operations for the same problem.

Step 3   Find the solution.
Use your math knowledge to find your answer.

Step 4   Check your answer.
Ask yourself, Is the answer reasonable? Did you find what you were asked for?

When solving word problems, sometimes you may need to find the areas of parts of figures.

Example

Refer to the figure below. Find the amount of waste as a result of cutting a semicircular plate from a piece of sheet metal that measures 9 inches by 4.5 inches.

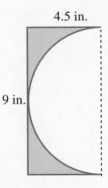

4.5 in.

9 in.

Step 1   The key words are **find the waste.**

Step 2   This problem asks for the amount of waste (the shaded area) created from cutting a piece of metal. To solve this problem, find the area of the rectangular piece of metal and then the area of the semicircular piece of metal. Then subtract the area of the semicircle from the area of the rectangle.

Step 3 The area of the rectangle is 9(4.5), or 40.5 square inches.
The area of the semicircle is

$$A = [3.14(4.5)(4.5)] \div 2$$
$$A = 63.585 \div 2$$
$$A = 31.7925$$

The waste is the difference between the areas of the rectangle and the semicircle.

$$40.5 - 31.7925 = 8.7075$$

The waste is 8.7075 square inches.

Step 4 The area of the rectangle is larger than that of the semicircle. The waste is small so the answer seems reasonable.

Practice

Use the following information to answer questions 1–3.

The distance around a circular flower bed is 50.24 feet. The flower bed is surrounded by a walk 3 feet wide.

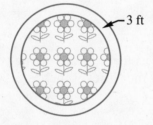

1. What is the diameter of the flower bed? _____

2. What is the area of the flower bed? _____

3. What is the area of the walk? _____

Use the following information to answer questions 4 and 5.

A race track is shaped like a rectangle with two semicircles on the shorter sides. The rectangular portion measures $\frac{3}{8}$ mile by $\frac{1}{8}$ mile.

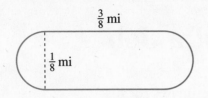

$\frac{3}{8}$ mi

$\frac{1}{8}$ mi

4. What is the area of the infield?   _____

5. What is the length of the track?   _____

Solve.

6. At $1.25 a square yard, how much would it cost to sod the playing field shown below?   _____

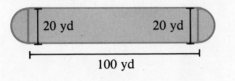

20 yd   20 yd

100 yd

7. Find the number of quarts of paint needed to paint 125 round concrete stepping-stones, each with a diameter of 2 feet. One quart covers 100 square feet.   _____

2 ft

Estimating the Cost of a Job

A contractor uses area formulas to compute estimates for his work. If he estimates the cost to be $20 per square yard, what price will the contractor quote to the park district for building a 1-yard-wide sidewalk around a circular flower bed 10 yards in diameter?

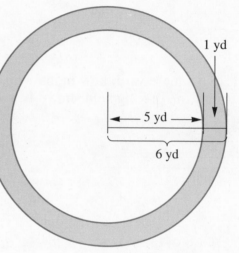

First, he finds the radius of the flower bed.

$$r = \frac{d}{2}$$

$$r = \frac{10 \text{ yd.}}{2} = 5 \text{ yd.}$$

Second, he computes the area of the flower bed.

$$A = \pi r^2$$
$$A = 3.14(5)(5)$$
$$A = 78.5 \text{ sq. yd.}$$

Third, he finds the radius of the flower bed plus the sidewalk.

$$r = 5 + 1 = 6 \text{ yd.}$$

Fourth, he computes the area of the flower bed plus the sidewalk.

$$A = \pi r^2$$
$$A = 3.14(6)(6)$$
$$A = 113.04 \text{ sq. yd.}$$

Fifth, he subtracts the area of the flower bed from the area of the flower bed plus the sidewalk to get the area of the sidewalk.

$$
\begin{array}{r}
113.04 \text{ sq. yd.} \\
-78.50 \\
\hline
34.54 \text{ sq. yd.}
\end{array}
$$

Sixth, he multiplies the area of the sidewalk by $20 to get the cost of the job.

$$
\begin{array}{r}
34.54 \\
\times \$20 \\
\hline
\$690.80
\end{array}
$$

1. At $1.75 a square foot, what price will be bid to pour a cement slab foundation for the housing project floor plan shown below?

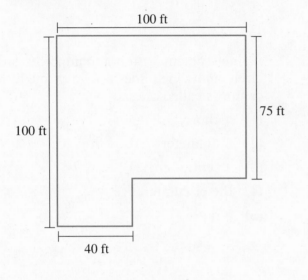

2. At $0.20 a square foot, what price will be quoted to pave the parking lot of this community center?

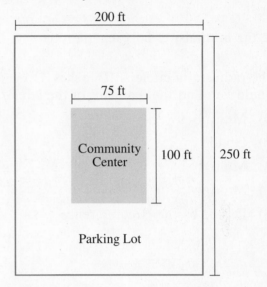

_____

_____

3. The Marco Company received bids from two contractors to tile the pavement around the company's outdoor reflecting pool.

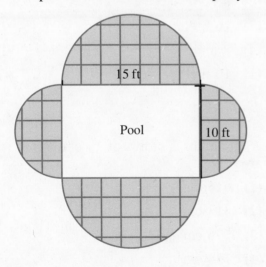

(1) Contractor A estimated ceramic tile at $1.49 per square foot and labor at $12 per hour for 15 hours. What was the bid?

_____

(2) Contractor B estimated the ceramic tile at $1.69 per square foot, and labor at $10.50 per hour for 20 hours. What was the bid?

_____

Posttest

Circle the best answer for each question.

1. The radius of a circle is 10 inches. How would you find the diameter of the circle?

   (1) Multiply by 2.

   (2) Divide by 2.

   (3) Multiply by π.

   (4) Multiply by 0.5.

   (5) Divide by the circumference.

2. A line segment with both endpoints on the circle and which does not go through the center is called _____ .

   (1) a chord

   (2) a diameter

   (3) a central angle

   (4) the circumference

   (5) a radius

3. The radius of a circle is _____ .

   (1) one-half the circumference ÷ diameter

   (2) equal to twice the diameter

   (3) one-half the circle

   (4) the distance around the circle

   (5) a straight line

4. The area of a circle is 50.24 square meters. What is its radius?

   (1) 4 meters

   (2) 8 meters

   (3) 16 meters

   (4) 2 meters

   (5) 10 meters

5. The radius of a circle is 14 cm. What is its circumference?

   (1) 22 cm

   (2) 44 cm

   (3) 87.92 cm

   (4) 88 cm

   (5) 616 cm

6. A circle has a radius of 14 cm. What is its area?

   (1) 616 cm^2

   (2) 313 cm^2

   (3) 196 cm^2

   (4) 154 cm^2

   (5) 87.92 cm^2

Problem Solving

139

Solve the following problems.

7. Allowing $\frac{1}{2}$ inch for the overlap, how long should a label be to go around a can with a diameter of 3 inches?

8. Find the length of a belt needed for two pulleys, 7 inches in diameter, whose centers are 24 inches apart. (Hint: The belt goes around only half of each pulley.)

7 in.   7 in.

_____   _____

Pretest

Use the grid below to answer the problems. Circle the best answer for each question.

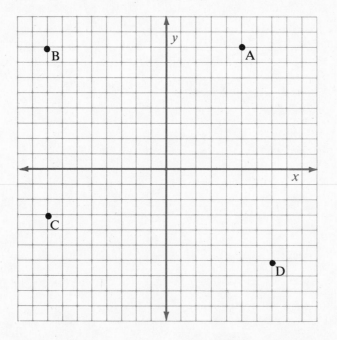

1. What are the coordinates of point *B*?

   (1) (8, 8)

   (2) (−8, −8)

   (3) (−8, 8)

   (4) (−9, 9)

   (5) (9, −9)

2. What are the coordinates of point *C*?

   (1) (−3, −8)

   (2) (−8, −3)

   (3) (−8, 3)

   (4) (8, −3)

   (5) (−4, −9)

3. To plot point E $(0, -9)$ on the graph, start at the origin and _____ .

   (1) count up 9 units

   (2) count down 9 units

   (3) count to the left 9 units

   (4) count to the right 9 units

   (5) count up 1 unit and to the left 9 units

4. To plot point F $(-2, -3)$ on the graph, start at the origin and _____ .

   (1) count down 2 and to the left 3 units

   (2) count to the left 5 units

   (3) count to the left 2 units and up 3 units

   (4) count to the right 2 units and down 3 units

   (5) count to the left 2 units and down 3 units

5. What is the distance between points A and B?

   (1) 3 units

   (2) -3 units

   (3) 13 units

   (4) -13 units

   (5) 19 units

6. What is the distance between points B and C?

   (1) 11 units

   (2) 5 units

   (3) 16 units

   (4) -16 units

   (5) -11 units

7. Use the formula

   $$AB = \sqrt{(x_2 - x_1)^2 + (y_2 - y_1)^2}$$

   to find the distance between point C and point D. The distance is about _____ .

   (1) 2 units

   (2) 17.5 units

   (3) 3.2 units

   (4) 15.3 units

   (5) 16 units

8. What is the midpoint of $\overline{CD}$?

   (1) $(0, 4)$

   (2) $(\frac{1}{2}, 4\frac{1}{2})$

   (3) $(-\frac{1}{2}, 4\frac{1}{2})$

   (4) $(-1, 5)$

   (5) $(-\frac{1}{2}, -4\frac{1}{2})$

9. What is the perimeter of triangle ABC?

   (1) 41 units

   (2) 37 units

   (3) 39 units

   (4) 56 units

   (5) 24 units

10. What is the area of triangle ABC?

    (1) 143 square units

    (2) 71.5 square units

    (3) 39.3 square units

    (4) 286 square units

    (5) 84.15 square units

Ordered Pairs and Graphing

A **plane** can be divided into four parts by a horizontal number line and a vertical number line. The horizontal number line is called the **x-axis.** The vertical number line is called the **y-axis.** They intersect in a point called the **origin** (labeled *0*).

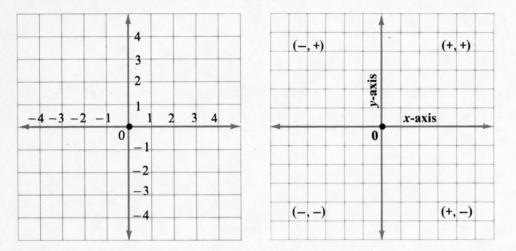

The location of any point in the plane can be identified by a pair of numbers called the **coordinates** (or **ordered pair**) of the point. The numbers are written inside parentheses.

(*x, y*) The first number tells the horizontal distance of the point from the origin. A positive number indicates that the point is to the right of the origin. A negative number indicates that the point is to the left of the origin.

The second number tells the vertical distance from the origin. A positive number indicates that the point is above the origin. A negative number indicates that the point is below the origin.

Examples

A. Locate and label point *A* (4, 0) and point *B* (−4, 0) on the grid.

For point *A*, start at the origin. Count to the right 4 units.

For point *B*, start at the origin. Count to the left 4 units.

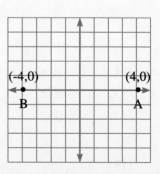

B. Identify the coordinates of each point on the grid.

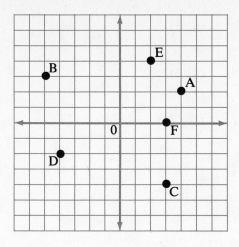

Point *A* is 4 units to the right of the origin and 2 units above the origin. The coordinates of *A* are (4, 2). The coordinates of point *B* are (−5, 3). The coordinates of the other points are as follows:

C is (3, −4).
D is (−4, −2).
E is (2, 4).
F is (3, 0).

Practice

Write the coordinates for each point shown on the grid.

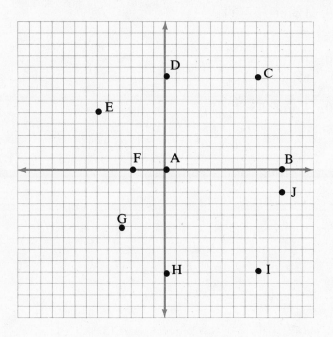

1. Point *A* = (_____ , _____)

2. Point *B* = (_____ , _____)

3. Point *C* = (_____ , _____)

4. Point *D* = (_____ , _____)

5. Point *E* = (_____ , _____)

6. Point *F* = (_____ , _____)

7. Point *G* = (_____ , _____)

8. Point *H* = (_____ , _____)

9. Point *I* = (_____ , _____)

10. Point *J* = (_____ , _____)

Locate and label each point on this grid.

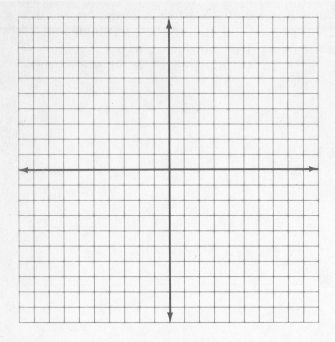

11. Point A = (0, 3)

12. Point B = (4, 4)

13. Point C = (−5, 0)

14. Point D = (−9, 1)

15. Point E = (−5, −5)

16. Point F = (−7, −2)

17. Point G = (0, −7)

18. Point H = (1, −1)

19. Point I = (3, −7)

20. Point J = (7, 0)

Distance

You can find the distance between two points on a graph if both points lie on a horizontal or vertical line.

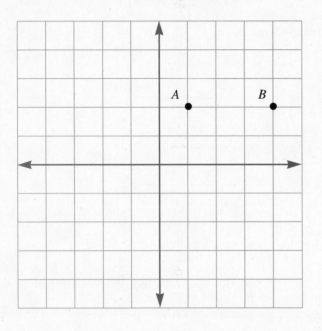

Count the units between points A and B. The distance between these points is 3 units.

You can also use the coordinates to find the distance. The ordered pair for point A is (1, 2).

The ordered pair for point B is (4, 2). Subtract the x-coordinates:

$$4 - 1 = 3$$

Subtract the y-coordinates:

$$2 - 2 = 0$$

Therefore, the distance between point A and point B is 3 units.

Example

A. Find the distance between points C and D.

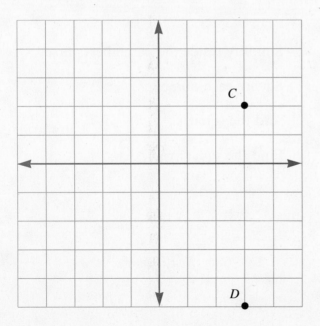

You can find the distance between the points by counting on the graph. You will find that it is 7 units.

You can also use the subtraction method to find the distance.

The ordered pair for point C is (3, 2). The ordered pair for point D is (3, −5).

$$3 - 3 = 0 \qquad \text{Subtract the } x\text{-coordinates.}$$
$$2 - (-5) = 2 + 5 = 7$$
$$\text{Subtract the } y\text{-coordinates.}$$

The distance between points C and D is 7 units.

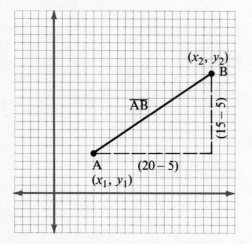
The distance between any point A and point B on a graph can be found by using the **Pythagorean Theorem.** $\overline{AB}$ is the hypotenuse of a right triangle. The legs are the x- and y-distances between points A and B.

Examples

B. Find the distance between point A and point B.

$$\overline{AB} = \sqrt{(x_2 - x_1)^2 + (y_2 - y_1)^2},$$

where point $A = (x_1, y_1)$
and point $B = (x_2, y_2)$.

$(y_2 - y_1)$ is the vertical distance between point A and point B.

$(x_2 - x_1)$ is the horizontal distance.

$\overline{AB} = \sqrt{(20 - 5)^2 + (15 - 5)^2}$ Simplify inside the parentheses first.

$\overline{AB} = \sqrt{15^2 + 10^2}$ Square each number.

$\overline{AB} = \sqrt{225 + 100}$ Add the squares.

$\overline{AB} = \sqrt{325}$ Find the square root.

$\overline{AB} = $ about 18 units

C. Find the distance between point *A* and point *B*.

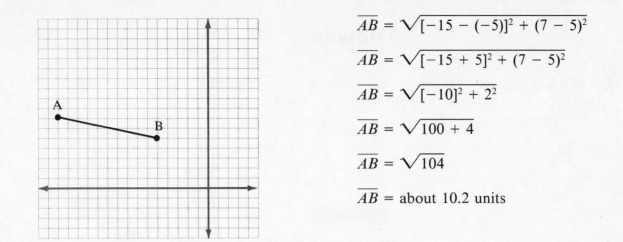

$$\overline{AB} = \sqrt{[-15 - (-5)]^2 + (7 - 5)^2}$$

$$\overline{AB} = \sqrt{[-15 + 5]^2 + (7 - 5)^2}$$

$$\overline{AB} = \sqrt{[-10]^2 + 2^2}$$

$$\overline{AB} = \sqrt{100 + 4}$$

$$\overline{AB} = \sqrt{104}$$

$$\overline{AB} = \text{about 10.2 units}$$

Practice

Find the distance between point *A* and point *B* for each of the following.

1. A(5, 2), B(5, −2) _____

2. A(5, 2), B(−2, 2) _____

3. A(0, 6), B(0, −5) _____

4. A(−2, −2), B(−7, −2) _____

Use the formula $\sqrt{(x_2 - x_1)^2 + (y_2 - y_1)^2}$ to find the distance between each set of points listed below.

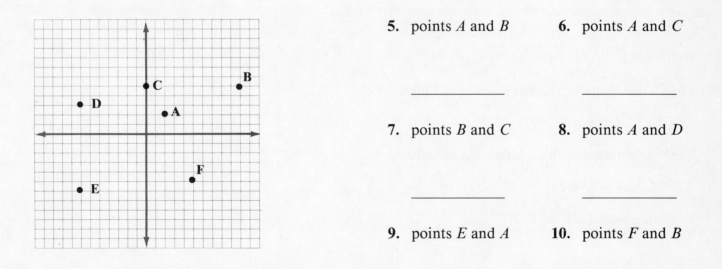

5. points *A* and *B*

6. points *A* and *C*

_____ _____

7. points *B* and *C*

8. points *A* and *D*

_____ _____

9. points *E* and *A*

10. points *F* and *B*

_____ _____

147

Midpoints

The **midpoint** of a line segment is the point that divides the line segment into two equal parts. The coordinates of the midpoint of a line segment are found by this formula:

$$\left(\frac{x_1 + x_2}{2}, \frac{y_1 + y_2}{2}\right)$$

Example

Find the midpoint of the line segment with endpoints A and B.

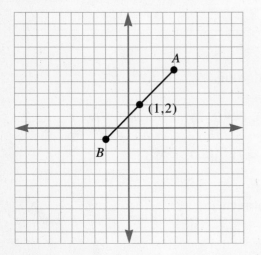

Point A is (4, 5).

Point B is $(-2, -1)$.

The x-coordinate of the midpoint is as follows:

$$\frac{4 + (-2)}{2} = \frac{2}{2} = 1$$

The y-coordinate of the midpoint is as follows:

$$\frac{5 + (-1)}{2} = \frac{4}{2} = 2$$

The midpoint of $\overline{AB}$ is (1, 2).

Use the midpoint formula to find the midpoint of each line segment. Use the graph to solve problems 1–6.

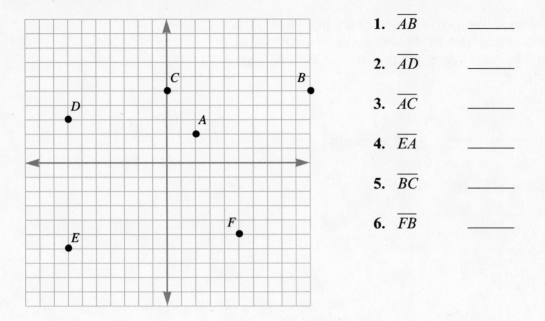

1. $\overline{AB}$ _____

2. $\overline{AD}$ _____

3. $\overline{AC}$ _____

4. $\overline{EA}$ _____

5. $\overline{BC}$ _____

6. $\overline{FB}$ _____

Problem Solving

Solve.

7. Find the midpoint of the line segment with endpoints $G(5, 2)$ and $H(15, 10)$. _____

8. Find the midpoint of the line segment with endpoints $I(-3, -2)$ and $J(-5, -2)$. _____

Perimeters and Areas of Plane Figures

By connecting the points on a graph, you can form a plane figure. You can determine the perimeter and area of a plane figure by first finding the lengths of the sides of the figure. Then apply the correct formula for finding area and perimeter.

Example

Find the perimeter and area of $\triangle ABC$.

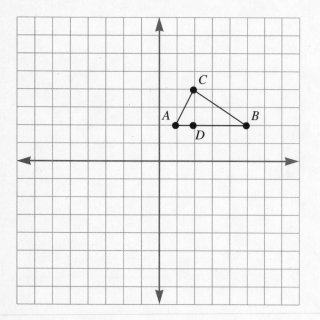

To find the perimeter of the triangle, determine the lengths of the sides of the triangle.

$$AB = 4 \text{ units}$$

Use the distance formula to find the lengths of AC and BC.

$$AC = \sqrt{(2 - 1)^2 + (4 - 2)^2}$$
$$= \sqrt{5}$$
$$= 2.236 \text{ units}$$
$$BC = \sqrt{(5 - 2)^2 + (2 - 4)^2}$$
$$= \sqrt{13}$$
$$= 3.606 \text{ units}$$

To find the perimeter, add the lengths of the sides together.

$$P = 4 + 2.236 + 3.606 = 9.842$$

The perimeter of $\triangle ABC$ is 9.842 units.

To find the area of $\triangle ABC$, use the formula, $A = \frac{1}{2}bh$. You must first determine the height of the triangle.

Use point C as the vertex for the height. Point D is the other endpoint of the line segment for the height. Point D has the ordered pair (2, 2). Therefore, $CD = 2$.

$A = \frac{1}{2}(AB)(CD)$

$A = \frac{1}{2}(4)(2)$

$A = 4$

The area of $\triangle ABC$ is 4 square units.

Graph the given points for each question. Then find the perimeter and area for each figure.

1. Connect points $A(4, 5)$, $B(-2, -1)$, and $C(-5, 5)$ to form a triangle.

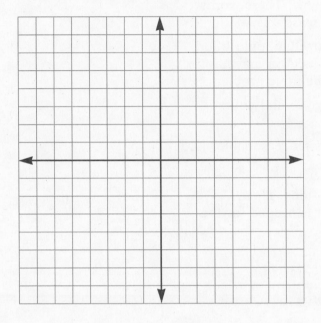

_____

_____

2. Connect points $W(-8, 7)$, $X(8, 7)$, $Y(8, -7)$. Find the fourth point needed to form a rectangle.

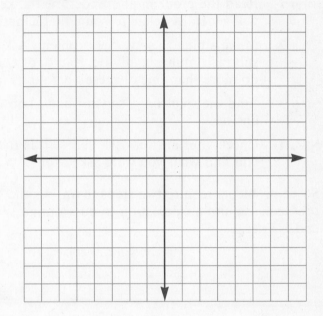

_____

_____

Problem Solving: Two-Step Problems

The steps you have learned to help solve word problems can be used with any word problems. Use the following steps:

Step 1   Read the problem and underline the key words. These words will generally relate to some mathematical reasoning.

Step 2   Make a plan to solve the problem. Ask yourself, Should I add, subtract, multiply, divide, round, or compare? You may have to do more than one of these operations for the same problem.

Step 3   Find the solution. Use your math knowledge to find your answer.

Step 4   Check your answer. Ask yourself, Is the answer reasonable? Did you find what you were asked for?

Some problems require two steps to answer the questions. These problems require a calculation to find a number that is needed for the end answer.

Example

The flat fee for renting a car is $25 a day plus $0.30 for every mile over 250 miles a day. Heidi rented a car for three days and traveled 224 miles, 325 miles, and 290 miles on each day, respectively. How much will she have to pay for renting the car?

Step 1   The key words are **how much, per day,** and **per mile.**

Step 2   To solve the problem, you need to figure out how many miles Heidi traveled. Then, subtract that total from the total allowed for three days. Multiply the difference in the miles by $0.30 and add the amount to the charge for three days.

Step 3   Solve.

$224 + 325 + 290 = 839$ miles traveled

Heidi is allowed $3(250)$, or 750 miles for the 3 days. She exceeded that amount by $839 - 750$, or 89 miles.

To determine how much Heidi paid for the excessive miles, multiply the excessive miles by $0.30.

$89 \times \$0.30 = 26.70$

Heidi paid $26.70 for the excessive miles.

Heidi paid $25 a day to rent the car. To determine the cost for renting the car for three days, multiply.

$$3 \times \$25 = \$75$$

Heidi paid $75 to rent the car for three days.

To find the total cost of renting the car, add the cost for the excessive miles to the flat rate.

$$\$26.70 + \$75.00 = \$101.70$$

The total cost of renting the car was $101.70.

Step 4 Check your answer.

$\$101.70 - \$75.00 = \$26.70$	This is the amount paid for the excessive miles.
$26.70 \div 0.30 = 89$	This is the number of miles over the allowed miles.
$89 + 750 = 839$	This is the number of miles traveled in the three days.
$839 = 224 + 325 + 290$	This checks.

The answer is reasonable.

Practice

Solve each of the following problems.

1. The fee for renting a big-screen television is $50, plus $5 per day. The store requires a deposit of $75 before a customer can take the television. If Linda rents the television for the Super Bowl and keeps it for five days, how much will she pay when she returns the set?

2. Mr. Wang is paid on commission. He receives $200 a week plus $45 for every refrigerator he sells. If he sells five refrigerators this week and three next week, how much pay will he receive for the two weeks?

3. Kyle runs 10 laps of a track every day. The track is circular with a diameter of 500 feet. How many miles does Kyle run in one week if he always runs on the circumference of the circle? (Hint: 1 mile = 5,280 feet)

_____

4. Ms. Smythe mows lawns during the summer. She mows four lots shaped like rectangles. The dimensions of the lots are 15 feet by 40 feet, 45 feet by 100 feet, 30 feet by 125 feet, and 35 feet by 50 feet. She charges $10 for the first 1,000 square feet and $1 for each additional 100 square feet. How much does she earn in one week?

_____

LIFE SKILL

Tracking Trends

Graphs can help track trends of investments in the stock market. By charting the trend of an investment over time, you can decide when to sell or buy stock.

Gloria bought a stock when it was worth $35 a share. Each week, she graphs the new price of the stock. The chart below represents her stock, and the graph shows the trend of the investment. Although the stock dropped significantly between weeks 2 and 3, Gloria's stock is increasing in value overall.

Week	Price	Week	Price	Week	Price
1	$35	6	$30	11	$40
2	$38	7	$32	12	$42
3	$20	8	$35	13	$38
4	$25	9	$39	14	$40
5	$25	10	$40	15	$45

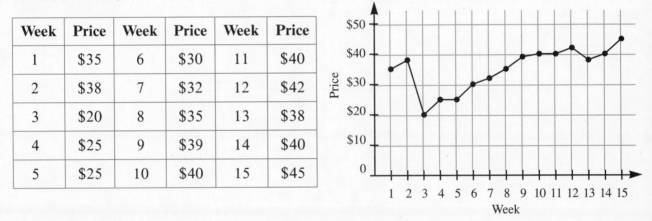

Charts can show other trends as well. For example, charting exercise repetitions, weight loss, or weight gain helps you see successes and weaknesses. Charting business trends, such as costs, helps a company to see a problem before it becomes too large to handle.

John is losing weight. He uses a chart of his weight to keep track of his success. Draw a chart on the graph below to show his weight over 10 weeks.

Week	Weight	Week	Weight
1	250 lbs	6	239 lbs
2	248 lbs	7	236 lbs
3	245 lbs	8	232 lbs
4	240 lbs	9	234 lbs
5	242 lbs	10	230 lbs

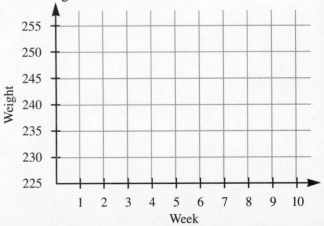

Posttest

Use the grid below to answer the problems. Circle the best answer for each question.

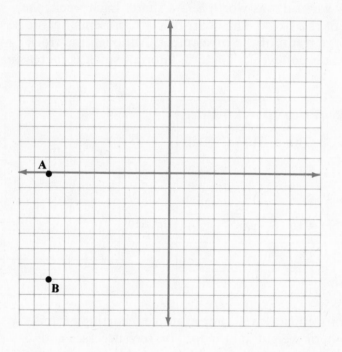

1. What are the coordinates of point *A*?

    (1) (0, 8)

    (2) (−8, 0)

    (3) (−8, 8)

    (4) (0, −8)

    (5) (−9, 0)

2. What are the coordinates of point *B*?

    (1) (−7, −8)

    (2) (7, −8)

    (3) (−8, 7)

    (4) (−8, −7)

    (5) (−9, −8)

3. To plot $C(-1, 0)$ on the graph, start at the origin and _____ .

   (1) count down 1 unit
   (2) count to the left 1 unit
   (3) count to the right 1 unit
   (4) count to the left 1 unit and down 1 unit
   (5) count down 1 unit and to the left 1 unit

4. To plot point $D(7, -6)$ on the graph, start at the origin and _____ .

   (1) count down 6 and to the left 7 units
   (2) count to the left 7 units and down 6 units
   (3) count to the left 7 units and up 6 units
   (4) count to the right 7 units and down 6 units
   (5) count to the right 7 units and up 6 units

5. What is the distance between points A and B?

   (1) -7 units
   (2) -16 units
   (3) -25 units
   (4) 25 units
   (5) 7 units

6. What is the distance between points A and C?

   (1) -9 units
   (2) 9 units
   (3) 7 units
   (4) -7 units
   (5) 8 units

7. Use the formula
   $$AB = \sqrt{(x_2 - x_1)^2 + (y_2 - y_1)^2}$$
   to find the distance between point B and point D. The distance is about _____ .

   (1) 15 units
   (2) $(-1, -13)$
   (3) 1.4 units
   (4) 13 units
   (5) 7.5 units

8. What is the midpoint of $\overline{BD}$?

   (1) $(-1, -13)$
   (2) $(\frac{1}{2}, 6\frac{1}{2})$
   (3) $(-\frac{1}{2}, 6\frac{1}{2})$
   (4) $(-\frac{1}{2}, -6\frac{1}{2})$
   (5) $(\frac{1}{2}, -6\frac{1}{2})$

9. What is the perimeter of triangle ABC?

   (1) 23.9 units
   (2) 29 units
   (3) 49 units
   (4) 21 units
   (5) 38.9 units

10. What is the area of triangle ABC?

   (1) 49 square units
   (2) 98 square units
   (3) 24.5 square units
   (4) 52.5 square units
   (5) 74.25 square units

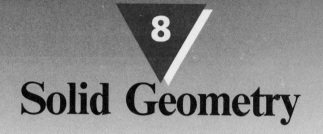

U N I T 8

Solid Geometry

────────── **Pretest** ──────────

Circle the best answer for each question.

1. Solid geometric figures have _____.
   (1) weight
   (2) height
   (3) width
   (4) depth
   (5) height, width, and depth

2. A rectangular solid has _____.
   (1) 6 faces
   (2) 8 faces
   (3) 12 faces
   (4) 6 edges
   (5) 3 edges

3. The surface area of a rectangular solid is _____.
   (1) the sum of the areas of all three faces
   (2) the product of the length, width, and height
   (3) the sum of the areas of all six faces
   (4) the product of the areas of the surfaces
   (5) equal to its volume

4. The volume of a rectangular solid is _____.
   (1) the sum of the areas of all six faces
   (2) the product of the length, width, and height
   (3) the product of the areas of the surfaces
   (4) equal to its surface area
   (5) measured in square units

5. A cube measures 4 feet on each side. What is the surface area of this cube?

    (1) 16 ft²

    (2) 32 ft²

    (3) 48 ft²

    (4) 64 ft²

    (5) 96 ft²

6. What is the volume of the cube in question 5?

    (1) 16 ft³

    (2) 32 ft³

    (3) 48 ft³

    (4) 64 ft³

    (5) 96 ft³

Use the figure below to answer questions 7 and 8.

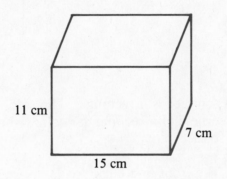

11 cm

7 cm

15 cm

7. What is the surface area?

    (1) 1,155 cm²

    (2) 347 cm²

    (3) 694 cm²

    (4) 270 cm²

    (5) 540 cm²

8. What is the volume?

    (1) 1,155 cm³

    (2) 347 cm³

    (3) 694 cm³

    (4) 270 cm³

    (5) 540 cm³

Problem Solving

9. A room measures 8 feet by 10 feet. The ceiling height is 8 feet. If the walls are to be covered with paneling, how much paneling is required?

    (1) 80 ft²

    (2) 288 ft²

    (3) 640 ft²

    (4) 448 ft²

    (5) 1280 ft²

10. A cement sidewalk will measure 3 feet wide, 20 feet long, and 6 inches deep. How many cubic feet of cement are required?

    (1) 396 cu ft

    (2) 360 cu ft

    (3) 120 cu ft

    (4) 30 cu ft

    (5) 29 cu ft

Rectangular Solids

Plane geometric figures, such as rectangles, triangles, and circles have only two dimensions. Solid geometry involves three dimensions. Solid geometric figures have **height, width,** and **length.**

A box is an example of a **rectangular solid.** It has three dimensions: length (*l*), width (*w*), and height (*h*).

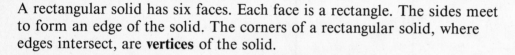

A rectangular solid has six faces. Each face is a rectangle. The sides meet to form an edge of the solid. The corners of a rectangular solid, where edges intersect, are **vertices** of the solid.

The top and bottom are rectangles with sides *l* and *w*.

The front and back are rectangles with sides *l* and *h*.

The right and left sides are rectangles with sides *w* and *h*.

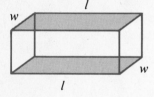
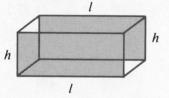
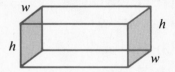

A diagonal of a rectangular solid is a line segment that joins two vertices that are not in the same plane.

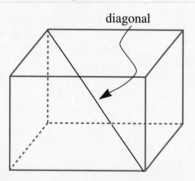

diagonal

If the length, width, and height of a rectangular solid are all equal, the solid is called a **cube.** The faces of the cube are squares. A sugar cube is an example of a cube.

Identify the parts of the rectangular solid.

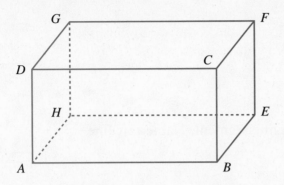

$\overline{AB}$, $\overline{BE}$, $\overline{EH}$, $\overline{AH}$, $\overline{DC}$, $\overline{CF}$, $\overline{FG}$, $\overline{DG}$, $\overline{AD}$, $\overline{BC}$, $\overline{HG}$, $\overline{EF}$ are edges.

Rectangles *ABCD*, *BEFC*, *EFGH*, *AHGD*, *ABEH*, *DCFG* are faces.

Use the figure below to answer questions 1–5.

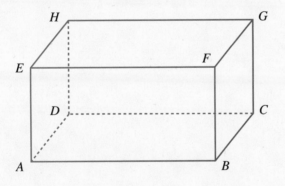

1. How many pairs of congruent, or equal, rectangles are contained by this rectangular solid? Name each congruent pair.

_____

_____

_____

_____

2. How many sets of equal edges are contained by this rectangular solid? Name each equal set.

_____

_____

_____

_____

3. Faces *ABCD* and *BCGF* intersect in a line segment. Name this line segment.

_____

4. What kind of angle is formed by the intersection of faces *ABCD* and *BCGF*?

_____

5. Name four diagonals of the rectangular solid.

_____

_____

_____

_____

Surface Areas of Rectangular Solids

The sum of the areas of the six faces of a rectangular solid is the **surface area** of the solid. Since all of the faces of a rectangular solid are rectangles, use the formula $A = lw$ to find the area of each face.

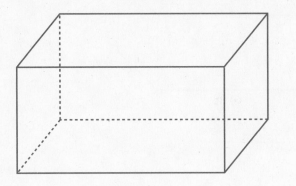

The top and bottom of the solid have equal areas. The front and back of the solid have equal areas. The left and right sides have equal areas. The formula for the surface area of a rectangular solid is:

$$S = 2lw + 2wh + 2lh$$

where S stands for surface area, l stands for length, w for width, and h for height.

Examples

A. Find the total surface area of this figure.

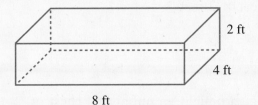

2 ft

4 ft

8 ft

$S = 2lw + 2wh + 2lh$
$S = 2(8)(4) + 2(4)(2) + 2(8)(2)$
$S = 64 + 16 + 32$
$S = 112$

The total surface area is 112 sq ft.

B. Find the amount of paint needed to paint the walls and ceiling of a room 12 feet long by 10 feet wide with a ceiling height of 8 feet.

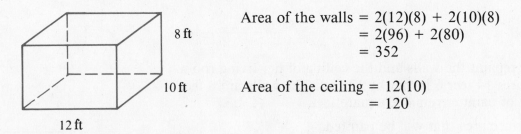

8 ft

10 ft

12 ft

Area of the walls = $2(12)(8) + 2(10)(8)$
= $2(96) + 2(80)$
= 352

Area of the ceiling = $12(10)$
= 120

The total area is 352 + 120, or 472 square feet, minus any area for doors and windows.

Solve.

1. Find the surface area of this box.

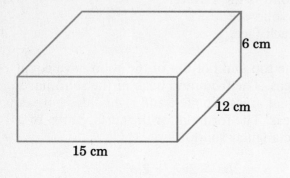

6 cm

12 cm

15 cm

_____

2. Find the surface area of this box.

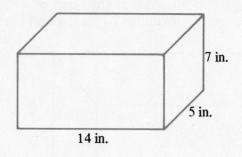

7 in.

5 in.

14 in.

_____

Problem Solving

Solve the following problems.

3. How many square inches of cardboard are needed to make a box 12 inches long, 8 inches wide, and 6 inches high?

_____

4. How much paper is needed to wrap a box 3 feet × 2 feet × 1½ feet?

_____

5. Chu wants to make a new bedspread for her bed. The bed measures 75 inches long, 54 inches wide, and 24 inches high. The spread is to cover the top, the sides, and the foot (but not the head) of the bed. How much fabric should Chu buy to make the spread?

_____

6. Miguel bought an unfinished chest of drawers. The chest is 48 inches wide, 30 inches tall, and 15 inches from front to back. Miguel will stain the top, the sides, and the front of the chest. How many square inches of the chest will Miguel stain?

_____

7. Cheleta plans to repaint the walls and the ceiling of her living room. The room measures 14 feet 6 inches by 12 feet. The walls are 8 feet high. One gallon of paint covers 400 square feet.

    (1) Find the surface area that will be painted.   _____

    (2) Find the amount of paint that is needed.   _____

LIFE SKILL

Paneling a Room

Bernie decided to panel his dining room. The room is 12 feet wide by 16 feet long by 10 feet high. How much paneling should he buy? He must buy the paneling in 4 × 8 foot sheets.

Step 1 Draw a model of the room. Draw the walls like the dropped sides of a box. Draw windows and doors.

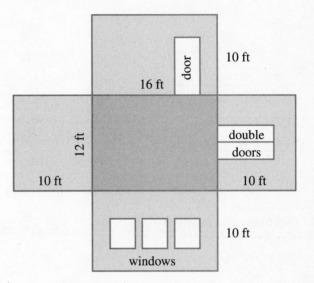

Step 2 Find the total area of the walls.
10 ft × 12 ft = 120 sq ft
10 ft × 12 ft = 120 sq ft
10 ft × 16 ft = 160 sq ft
10 ft × 16 ft = <u>160 sq ft</u>
Total = 560 sq ft

Step 3 Divide the total area by the area of one panel.
560 ÷ 32 = $17\frac{1}{2}$ panels

Step 4 Subtract $\frac{1}{2}$ panel for each window.

Subtract $\frac{2}{3}$ panel for each door.

For windows: $\frac{1}{2} \times \frac{3}{1} = \frac{3}{2} = 1\frac{1}{2}$ $1\frac{1}{2}$ $17\frac{1}{2}$ panels

For doors: $\frac{2}{3} \times \frac{3}{1} = \frac{6}{3} = 2$ $\underline{+\ 2}$ $\underline{-\ 3\frac{1}{2}}$ panels
                                                    $3\frac{1}{2}$ 14 panels

Bernie needs to buy 14 panels for the room.

How many sheets of paneling are needed for these rooms?

1. Room dimensions: 10 ft × 15 ft × 8 ft
   Windows: 2   Doors: 1   _____

2. Room dimensions: $12\frac{1}{2}$ ft × $14\frac{3}{4}$ ft × 8 ft
   Windows: 4   Doors: 1 double, 2 single   _____

3. Room dimensions: $18\frac{2}{3}$ ft × $11\frac{2}{3}$ ft × 8 ft
   Windows: 5   Doors: 2 doubles, 1 single   _____

165

Volumes of Rectangular Solids

Volume is a measure of capacity. This is how much an object can hold or how much space it takes up.

Volume is measured in cubic units. A cubic centimeter is a cube that measures one centimeter on each side. To find the number of cubic centimeters contained in the rectangular solid below, you can divide the solid into layers as shown.

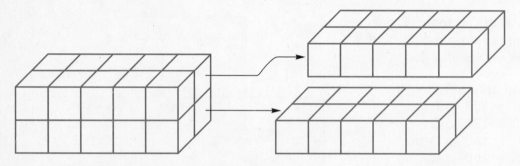

The bottom layer is the base. The area of the base of the rectangular solid is the length times the width, and the number of layers is the height.

The formula for finding the volume (V) of a rectangular solid is

$$V = lwh$$

where V stands for volume, l stands for length, w stands for width, and h stands for height.

Examples

A. Find the volume of the figure below.

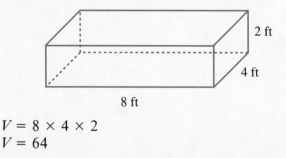

$V = 8 \times 4 \times 2$
$V = 64$

The volume of the figure is 64 cubic feet (ft^3).

B. How much cement does a concrete patio 15 feet long, 11 feet wide, and 4 inches thick require?

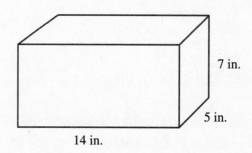
First change the inches to $\frac{1}{3}$ foot.
Use the volume formula.

$$V = lwh$$
$$V = 15(11)(\tfrac{1}{3})$$
$$V = 55$$

The patio requires 55 cubic feet of concrete.

Practice

Solve.

1. Find the volume of the box shown below.

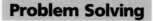

6 cm

12 cm

15 cm

2. Find the volume of the figure shown below.

7 in.

5 in.

14 in.

_____   _____

Problem Solving

Solve the following problems.

3. A cubic foot measures one foot on each side. Find the volume of a cubic foot in cubic inches.

4. A cubic yard measures one yard on each side. Find the volume of a cubic yard in cubic feet.

_____   _____

5. What is the volume of a box that is 12 inches long, 8 inches wide, and 6 inches high?

_____

6. How much dirt is removed for a basement 60 feet long, 25 feet wide, and 8 feet deep?

_____

7. How much topsoil is needed to cover a lot 110 feet by 90 feet to a depth of 4 inches?

_____

8. One cubic foot of cast brass weighs about 525 pounds. How much does a bar of cast brass 20 inches long, 15 inches wide, and 12 inches thick weigh?

_____

9. An aquarium is 2 feet long, 1 foot wide, and 1 foot tall.

  (1) How many gallons of water will it hold?
  (Hint: 1 cubic foot = 7.48 gallons)

_____

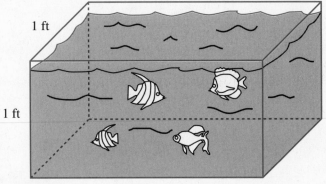

1 ft

1 ft

2 ft

  (2) A gallon of water weighs about 8.3 pounds. How much will the aquarium weigh when it is filled with water?

_____

Problem Solving: Choosing the Correct Formula

The steps you have learned to help solve word problems can be used with word problems that deal with geometric figures. Use the following steps:

Step 1   Read the problem and underline the key words. These words will generally relate to some mathematical reasoning.

Step 2   Make a plan to solve the problem. Ask yourself, Should I add, subtract, multiply, divide, round, or compare? You may have to do more than one of these operations for the same problem.

Step 3   Find the solution. Use your math knowledge to find your answer.

Step 4   Check your answer. Ask yourself, Is the answer reasonable? Did you find what you were asked for?

When working with rectangular solids, you must decide whether you need to find the surface area or the volume of a figure. If the problem involves surface area, you must further decide whether you need the total area or only some of the faces.

Examples

A.   A factory packs certain automobile parts in boxes that measure 3 in. × 5 in. × 2 in. How many of these boxes fit into packing crates measuring 36 in. × 20 in. × 24 in.?

Step 1   The key words are **capacity** and **how many.**

Step 2   First find the volume of a box for individual parts. Then, find the volume of the crate. To find the number of boxes that will fit in the crate, divide the volume of the crate by the volume of an individual box.

Step 3   To find the volume of a box, use the formula $V = lwh$.

$3 \times 5 \times 2 = 30$

The volume of an individual box is 30 cu in.

To find the volume of a crate, you will again use the formula $V = lwh$.

$36 \times 20 \times 24 = 17{,}280$

The volume of a crate is 17,280 cu in.

To find out how many boxes the crate will hold, divide the volume of the crate by the volume of an individual box.

$$17{,}280 \div 30 = 576$$

The crate will hold 576 boxes.

Step 4 To check your answer, multiply.

$$30 \times 576 = 17{,}280$$

This checks. The answer is reasonable.

B. How much paper do you need to wrap a present that measures 10 in. × 8 in. × 6 in.?

Step 1 The key words are **wrap a present** and **how much.**

Step 2 Since you are covering the surface of a present, you need to use surface area. Surface area needs to be found for all six faces.

Step 3 To find the solution, use the following formula:

$$S = 2lw + 2wh + 2lh$$
$$S = 2(10)(8) + 2(8)(6) + 2(10)(6)$$
$$S = 160 + 96 + 120$$
$$S = 376$$

The surface area of the present is 376 sq in.

Step 4 Check your answer.

$$S = 10(8) + 8(6) + 10(6) + 10(8) + 8(6) + 10(6)$$
$$S = 376$$

The answer checks.

Answer each of the following.

1. A manufacturer wants to make a cardboard box that has a volume of 64 cubic centimeters. Which of these dimensions could the manufacturer use? _____

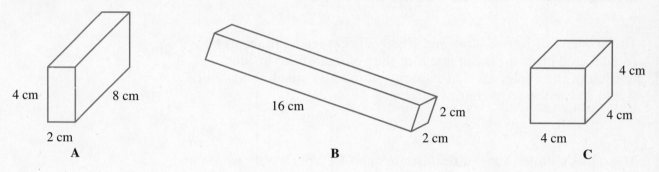

2. Which box would require the least amount of cardboard?

_____

3. Can two boxes have the same volume but different surface areas?

_____

4. During a rainfall of $1\frac{1}{2}$ inches, how much water would fall on a square foot of ground?

_____

5. The bed of a dumptruck is 15 feet long, 7 feet wide, and 6 feet tall. How much dirt will the truck hold?

_____

Capacity of Refrigerators

The capacity of refrigerators and freezers is expressed in terms of cubic feet. This is a measurement of the volume of the inside portion of the appliance. The capacity shows how much food can be stored inside the refrigerator or freezer.

The outside dimensions of refrigerators and freezers are shown on a sticker on the inside of the doors. The refrigerator shown is $66\frac{5}{8}$ inches in height, $35\frac{3}{4}$ inches wide, and $31\frac{1}{4}$ inches deep, and its capacity is 23.5 cubic feet.

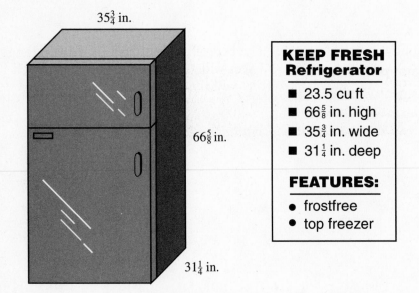

$35\frac{3}{4}$ in.

$66\frac{5}{8}$ in.

$31\frac{1}{4}$ in.

KEEP FRESH Refrigerator

- 23.5 cu ft
- $66\frac{5}{8}$ in. high
- $35\frac{3}{4}$ in. wide
- $31\frac{1}{4}$ in. deep

FEATURES:

- frostfree
- top freezer

Use the dimensions of the refrigerator to find its total volume. Compare this figure to the capacity shown on the sticker. How much of the refrigerator is needed for coolant and machinery?

To find the volume of the refrigerator, use the formula $V = lwh$.

$V = 66\frac{5}{8}(35\frac{3}{4})(31\frac{1}{4})$
$V = 74{,}433$ cubic inches

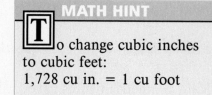

Change the cubic inches to cubic feet.

$74{,}433 \div 1{,}728 = 43.07$ cubic feet of capacity

To find out how much space is needed for coolant and machinery, subtract the capacity volume on the sticker (23.5) from the total volume of the refrigerator (43.07).

$43.07 - 23.5 = 19.57$

The coolant and machinery need 19.57 cubic feet of space.

Find the capacity of a refrigerator 60 inches tall by 30 inches wide by 24 inches deep. If the machinery takes about half of the total capacity, what is the space allotted for food storage?
Express the answer in feet.

_____

Posttest

Circle the best answer for each question.

1. Solid geometric figures have _____ .
   (1) only two dimensions
   (2) have three dimensions
   (3) no depth
   (4) two more dimensions than plane geometric figures
   (5) no volume

2. How many edges does a rectangular solid have?
   (1) 4 edges
   (2) 6 edges
   (3) 8 edges
   (4) 10 edges
   (5) 12 edges

3. Opposite faces of a rectangular solid _____ .
   (1) are perpendicular
   (2) intersect in a straight line
   (3) are congruent rectangles
   (4) have a common vertex
   (5) have equal volume

4. What is the formula for finding the surface area of a rectangular solid?
   (1) $S = 2l + 2w$
   (2) $S = 2lw + 2wh + 2lh$
   (3) $S = lw + wh + lw$
   (4) $S = lwh$
   (5) $S = 2l + 2w + 2h$

5. One side of a cube measures 3 in. What is the surface area of this cube?
   (1) 27 in^2
   (2) 36 in^2
   (3) 54 in^2
   (4) 108 in^2
   (5) Cannot be determined from this information.

6. What is the volume of the cube described in question 5?
   (1) 9 in^3
   (2) 27 in^3
   (3) 18 in^3
   (4) 54 in^3
   (5) Cannot be determined from this information.

Use the figure below to answer questions 7 and 8.

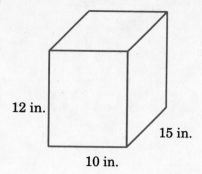

12 in.

15 in.

10 in.

7. What is the surface area of the figure?

   (1) 1,800 in²

   (2) 1,800 in³

   (3) 450 in²

   (4) 900 in²

   (5) 33 in²

8. What is its volume?

   (1) 1,800 in²

   (2) 1,800 in³

   (3) 450 in³

   (4) 900 in³

   (5) 33 in³

Problem Solving

9. A room measures 8 feet by 10 feet. The ceiling height is 8 feet. If the walls and the ceiling are to be painted, what is the size of the area requiring paint?

   (1) 144 ft²

   (2) 288 ft²

   (3) 368 ft²

   (4) 448 ft²

   (5) 352 ft²

10. A cement sidewalk will measure 1 meter wide, 7 meters long, and 10 centimeters deep. How many cubic meters of cement are required?

   (1) 70 cubic meters

   (2) 700 cubic meters

   (3) 7 cubic meters

   (4) 0.07 cubic meters

   (5) 0.7 cubic meters

9

Cylinders and Cones

Pretest

Circle the best answer for each question.

1. A cylinder has _____ .

   (1) two circular bases

   (2) one circular base and a vertex

   (3) a slant height

   (4) three faces

   (5) no flat surfaces

Use this figure to answer questions 2 and 3.

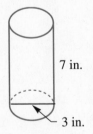

7 in.

3 in.

2. What is the total surface area?

   (1) $49\frac{1}{2}$ in^2 **(2)** 462 in^2

   (3) $14\frac{1}{7}$ in^2 **(4)** $80\frac{1}{7}$ in^2

   (5) 21 in^2

3. What is the volume?

   (1) $49\frac{1}{2}$ in^3 **(2)** 462 in^3

   (3) $14\frac{1}{7}$ in^3 **(4)** $476\frac{1}{7}$ in^3

   (5) 21 in^3

Use this figure to answer questions 4 and 5.

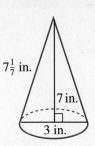

7\frac{1}{7} in.

7 in.

3 in.

4. What is the total surface area?
   (1) $14\frac{1}{7}$ in²
   (2) $34\frac{4}{49}$ in²
   (3) $40\frac{73}{98}$ in²
   (4) 21 in²
   (5) 150 in²

5. What is the volume?
   (1) $16\frac{1}{2}$ in³
   (2) $49\frac{1}{2}$ in³
   (3) 150 in³
   (4) 50 in³
   (5) 450 in³

6. Find the volume of this figure. (Hint: This is half of a cylinder over a rectangular solid.)

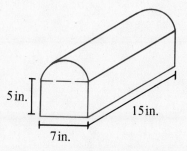

5 in.

15 in.

7 in.

   (1) 525 in³
   (2) 1,155 in³
   (3) $577\frac{1}{2}$ in³
   (4) $813\frac{3}{4}$ in³
   (5) 1,680 in³

7. Find the volume of this figure. (Hint: This is a rectangular solid with a cylinder removed.)

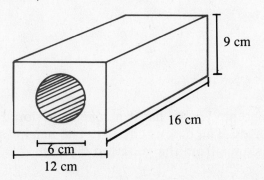

9 cm

16 cm

6 cm

12 cm

   (1) 452.16 cm³
   (2) 1,728 cm³
   (3) 1,275.84 cm³
   (4) 1,017.36 cm³
   (5) 2,745.36 cm³

Parts of Cylinders and Cones

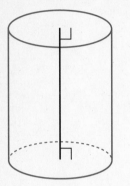

A cylinder has two circular bases. Each base is the same size. The radius of the bases is the radius of the cylinder. The segment joining the centers of the bases is called the **axis.** If the axis is perpendicular to the bases, the cylinder is a **right circular cylinder.** The figure shows a right circular cylinder. If the axis is not perpendicular to the bases, the cylinder is called an **oblique circular cylinder.** The height of a cylinder is the distance between its bases. In a right circular cylinder, the axis is the height.

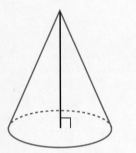

This is a cone. The base is a circle. The radius of the base is the radius of the circle.

The height of the cone is the distance between the vertex and the base.

Examples

A. Identify the parts of a soup can.

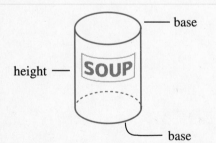

A soup can is a cylinder. It has two circular bases. The height of the can is the width of the label.

B. Identify the parts of a pointed party hat.

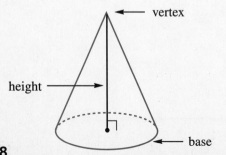

A pointed party hat is a cone. The point is the vertex. The base is circular. The height is the distance from the vertex to the base.

Solve each of the following.

1. Give four examples of cylinders in the classroom.

    _____

    _____

    _____

2. Give four examples of cones.

    _____

    _____

    _____

Label the parts of each figure.

3.

(b) _____

(c) _____

(a) _____

4.

(b) _____

(c) _____

(a) _____

Surface Areas of Cylinders and Cones

The **surface area** of a right circular cylinder is the sum of the areas of the two bases and the area of the curved surface. The area of the curved surface is called the **lateral surface area.** On a soup can, the lateral surface area is the part of a can covered by a label.

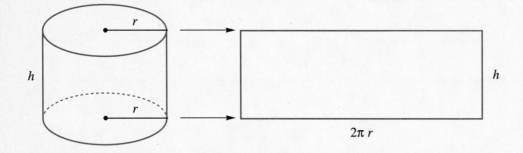

If you would cut the cylinder open and lay it flat, you would find that the lateral surface area is the area of a rectangle. This rectangle has a length equal to the circumference of the base of the cylinder. The height of the rectangle would be equal to the height of the cylinder.

To find the lateral surface area of a cylinder use the following formula:

$$LA = 2\pi rh$$

where LA is lateral surface area, r is the radius, and h is the height.

To find the areas of the bases of the cylinder, use the formula:

$$A = 2\pi r^2$$

where A is the areas and r is the radius.

To find the total surface area of a right circular cylinder, use the formula:

$$SA = 2\pi rh + 2\pi r^2$$

A. Find the total surface area of this right circular cylinder. Use $\frac{22}{7}$ for π.

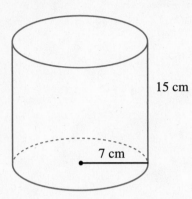

15 cm

7 cm

Area of the bases $= 2\pi r^2$

$$A = 2\left(\frac{22}{7}\right)(7)(7)$$

The area of the bases is 308 cm².

Lateral surface area $= 2\pi rh$

$$LA = 2\left(\frac{22}{7}\right)(7)(15)$$

The lateral surface area is 660 cm².

The total surface area is 308 + 660, or 968 cm².

The surface area of a right cone is the sum of the area of the base and the area of the curved surface. The area of the curved surface is the lateral surface area.

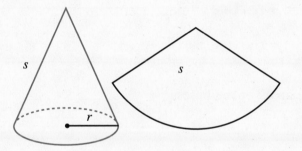

If you cut open a cone and lay it flat, you find the lateral surface area is the area of a part of a circle. To find the lateral surface area of a cone, use the following formula:

$$LA = \pi rs$$

where LA is the surface area, r is the radius of the base, and s is the distance from the vertex to the base. This is the height of the cone.

Use the following formula to find the area of the base of the circle:

$$A = \pi r^2$$

To find the total surface area of a cone, add the area of the lateral surface and the area of the base:

$$SA = \pi rs + \pi r^2$$

B. Find the total surface area of this right cone.

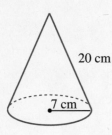

20 cm

7 cm

Area of the base $= \pi r^2$

$A = \left(\frac{22}{7}\right)(7)(7)$

$A = 154$

The area of the base is 154 cm².

Lateral surface area $= \pi rs$

$SA = \left(\frac{22}{7}\right)(7)(20)$

$SA = 440$

The lateral surface area is 440 cm².

The total surface area is 154 + 440, or 594 cm².

Practice

Use this figure to answer questions 1–3.

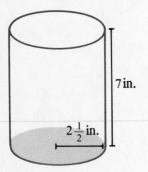

7 in.

$2\frac{1}{2}$ in.

1. Find the areas of the bases. _____

2. Find the lateral surface area. _____

3. Find the total surface area. _____

Use this figure to answer questions 4–6.

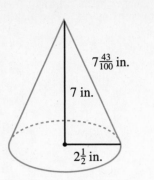

4. Find the area of the base. _____

5. Find the lateral surface area. _____

6. Find the total surface area. _____

For problems 7 and 8, find the total surface area.

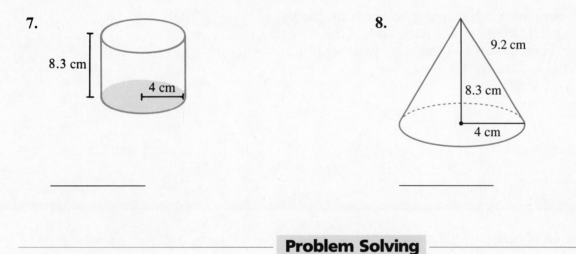

7. _____

8. _____

_____ **Problem Solving** _____

Solve.

9. How much metal is needed to make a 4-foot diameter pipe that is 8 feet long?

_____

10. Some people wrap water heaters to keep in the heat. If a water heater is 1.5 feet in diameter and 5 feet tall, how much wrap is needed?

_____

11. An oil refinery storage tank is in the shape of a cylinder. It is 50 feet tall and has a diameter of 120 feet. If the tank is to be painted, what is the size of the area requiring paint?

_____

12. A building has a roof in the shape of a right circular cone. The height of the roof is 12 feet and the diameter of the base is 10 feet. What is the lateral area of the roof?

_____

Sewing

Thomas is making a Halloween costume. For the costume, he wants to make a pointed hat shaped like a cone. The hat is to have a 12-inch slant height and a 7-inch diameter. The hat also needs a brim that is 4 inches wide.

In order to find out how much material he needs for the hat, Thomas must calculate the lateral surface area of the cone and the area of the brim. Then he can add the areas together to find the total amount of material needed for the hat.

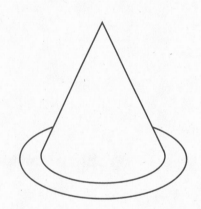

Lateral surface area of cone

Lateral surface area = $\pi r s$

$$LA = \left(\tfrac{22}{7}\right)\left(\tfrac{7}{2}\right)(12)$$
$$LA = 132$$

The lateral surface area of the cone is 132 sq in.

The Area of the Brim

To find the area of the brim, first find the areas of two circles. To find the area of the circle with a 7-inch diameter, use the formula $A = \pi r^2$.

$$A = \overset{11}{\underset{1}{\tfrac{22}{7}}} \times \overset{1}{\underset{1}{\tfrac{7}{2}}} \times \tfrac{7}{2}$$
$$A = \tfrac{77}{2}$$
$$A = 38\tfrac{1}{2}$$

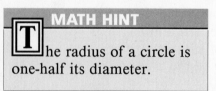

MATH HINT

The radius of a circle is one-half its diameter.

The area of the circle with a 7-inch diameter is $38\tfrac{1}{2}$ sq in.

To find the area of the circle with an 11-inch diameter, do the following:

$$A = \frac{\overset{11}{\cancel{22}}}{7} \times \frac{11}{\underset{1}{\cancel{2}}} \times \frac{11}{2}$$

$$A = \frac{1331}{14}$$

$$A = 95\frac{1}{14}$$

The area of the circle with an 11-inch diameter is $95\frac{1}{14}$ sq in.

Now subtract the area of the smaller circle from the area of the larger circle to find the area of the brim:

$$95\frac{1}{14} - 38\frac{1}{2} = 56\frac{6}{7}$$

The area of the brim is $56\frac{6}{7}$ sq in.

To find the total amount of material needed for the hat, add the surface area of the hat and the area of the brim.

$$132 + 56\frac{6}{7} = 188\frac{6}{7}$$

The total amount of material needed is $188\frac{6}{7}$ sq in.

Sylvia plans to make the skirt shown below. How much material does she need? The top part, if extended, would make a cone that would have a slant height of 45 inches.

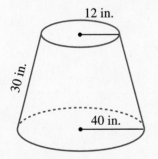

_____

Volumes of Cylinders and Cones

The volume of a cylinder is equal to the area of the base (πr^2) multiplied by the height (h). The formula for the volume of a cylinder is:

$$V = \pi r^2 h$$

The volume of a right circular cone is one-third the area of the base multiplied by the height. The formula for the volume of a right cone is:

$$V = \tfrac{1}{3}\pi r^2 h$$

Examples

A. Find the volume of this right cylinder.

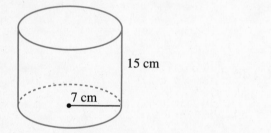

15 cm

7 cm

$V = \pi r^2 h$

$V = \dfrac{22}{7} \times \dfrac{\overset{1}{7}}{1} \times \dfrac{7}{1} \times \dfrac{15}{1}$

$V = 154$

$V = (154)(15)$

$V = 2{,}310$

The volume of the cylinder is 2,310 cm.

B. Find the volume of this right cone.

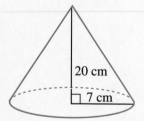

20 cm

7 cm

$V = \dfrac{22}{7} \times \dfrac{\overset{1}{7}}{1} \times \dfrac{7}{1} \times \dfrac{20}{1}$

$V = 154$

$V = \tfrac{1}{3}\pi r^2 h$

$V = \tfrac{1}{3}(154)(20)$

$V = \tfrac{1}{3}(3{,}080)$

$V = 1{,}026\tfrac{2}{3}$

The volume of the cone is about $1{,}026\tfrac{2}{3}$ cm³.

C. A grain silo in the shape of a cylinder has a radius of 9 yards and is 70 feet tall. How many bushels of grain can be stored in the silo? Use 3.14 for π. (One bushel takes up almost 1.25 cubic feet.)

$V = \pi r^2 h$ Change 9 yards to 27 feet.
$V = 3.14(27)^2(70)$
$V = 3.14(729)(70)$
$V = 160{,}234.2$

The volume of the silo is 160,234.2 cubic feet.

$160{,}234.2 \div 1.25 = 128{,}187.36$

The silo holds approximately 128,187 bushels of grain.

Practice

Find the volume.

1.

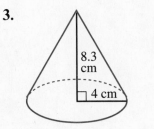

7 in.

$2\frac{1}{2}$ in.

_____

2.

7 in.

$2\frac{1}{2}$ in.

_____

3.

8.3 cm

4 cm

_____

4.

8.3 cm

4 cm

_____

Solve the following problems.

5. What is the volume of 8-foot long pipe with a 4-foot diameter?

    _____

6. How much dirt must be removed from a hole to make a well with a diameter of 5 feet and a depth of 49 feet?

    _____

7. A storage tank is in the shape of a cylinder. It is 50 feet tall and has a diameter of 120 feet. What is the capacity of the tank? (1 cubic foot = 7.48 gallons)

    _____

8. If a water heater is 1.5 feet in diameter and 5 feet tall, what is its capacity?

    _____

9. A pile of grain inside a silo is in the shape of a cone. The cone is 14 feet in diameter and 18 feet tall. How many bushels of grain are in this pile? (1 bushel ≈ 1.25 cu ft)

    _____

10. A tower has a roof in the shape of a right circular cone. The height of the roof is 12 feet and the diameter of the base is 10 feet. What is the volume of the roof?

    _____

LIFE SKILL

Cylinder of an Engine

The length of the diameter of a cylinder in an automobile engine is the **bore.** The distance the piston moves in the cylinder is the **stroke.** The engine capacity, or **displacement,** of a car is the combined volume of all its cylinders.

To find the displacement of a piston, find the volume of the cylinder.

Use the formula

$$V = \pi r^2 h$$

where V is the volume, r is the radius, and h is the height (or stroke).

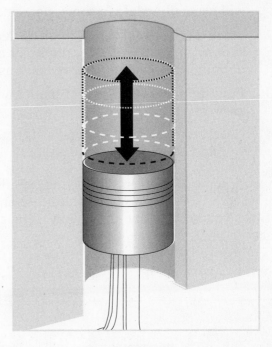

Find the displacement of a piston with a 4-inch bore and a 5-inch stroke.

$$V = \pi r^2 h$$
$$V = 3.14(2)^2 5$$
$$V = 62.8$$

MATH HINT

The bore measurement is a diameter, so you need to divide it by 2 to find the radius.

The displacement of the piston is 62.8 cubic inches.

Answer each of the following questions.

1. Find the total displacement for an engine with 8 cylinders using the specifications of the example.

2. Find the displacement of a 6-cylinder engine with a bore of 3.88 inches and a stroke of 3.25 inches.

_____

_____

Problem Solving: Deciding Which Formula to Use

The steps you have learned to help solve word problems can be used with word problems that deal with geometric figures. Use the following steps:

Step 1 Read the problem and underline the key words. These words will generally relate to some mathematical reasoning.

Step 2 Make a plan to solve the problem. Ask yourself, Should I add, subtract, multiply, divide, round, or compare? You may have to do more than one of these operations for the same problem.

Step 3 Find the solution. Use your math knowledge to find your answer.

Step 4 Check your answer. Ask yourself, Is the answer reasonable? Did you find what you were asked for?

When working with problems involving cylinders and cones, you must decide whether you need to find the surface area or the volume. Sometimes you need to find the volume of parts. To do this, add parts or subtract the parts from the whole.

Example

Find the volume of this metal block. There are two cylinders with diameters of 1-inch drilled out of the block.

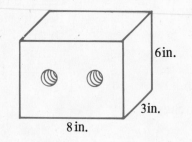

Step 1 The key word is **volume.**

Step 2 To solve this problem, find the volumes of the block and of the cylinders. Then subtract the volumes of the cylinders from the total volume of the block.

Step 3   To find the volume of the block, use the formula $V = lwh$.

$$V = 8(3)(6)$$
$$V = 144$$

The volume of the block is 144 cu in.

To find the volume of a cylinder, use this formula:

$$V = \pi r^2 h$$
$$V = 3.14(0.5)^2 (3)$$
$$V = 3.14(0.25)(3)$$
$$V = 0.785(3)$$
$$V = 2.355$$

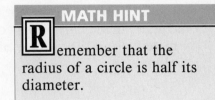

MATH HINT

Remember that the radius of a circle is half its diameter.

The volume of one cylinder is 2.355 cu in.

To find the volume of the two cylinders, multiply.

$$V = 2.355 \times 2 = 4.71$$

The total volume of the two cylinders is 4.71 cu in.

To find the volume of the block minus the cylinders, subtract.

$$144 - 4.71 = 139.29$$

The volume of the block minus the cylinder is 139.29 cu in.

Step 4   Check your answer by asking yourself, Does the answer seem reasonable?

The problem wants you to find volume. The formula for the volume of a cylinder is $V = lwh$. The cylinders cut out are very small, so the volumes of the block with the hole and without the holes should be close.

The answer seems reasonable.

Solve.

1. The base of a statue is shown below. The diameter of the bottom of the cone is 2 meters and the diameter of the top is 1 meter. The height of the base of the statue is 2 meters. If the entire cone were included, the height of the cone would be 4 meters. Find the volume of the base of the statue.

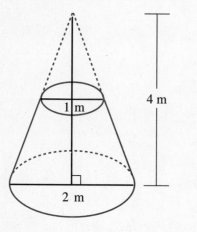

2. A greenhouse is in the shape of one-half of a right circular cylinder. It is 25 ft in diameter and 45 ft long. Except for the semicircular frames, the entire structure is translucent (clear) in order to provide the plants with the greatest possible exposure to sunlight. How much translucent material is needed to cover the semicircular frames of the greenhouse?

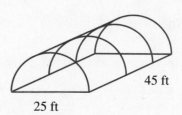

3. Georgio's driveway must cross a ditch that carries rainwater. He decides to use a 4-foot-diameter pipe to carry the water. He will pour a concrete driveway around it. Using the measurements on the diagram below, figure how many cubic yards of concrete he will need?
Hint: 27 cubic feet = 1 cubic yard.

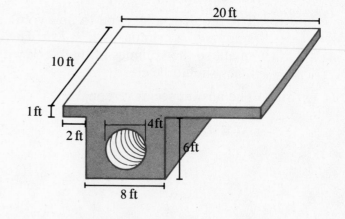

Posttest

Circle the best answer for each question.

1. A cone has ——————— .
   (1) two circular bases
   (2) one circular base and a vertex
   (3) a slant height that is equal to its height
   (4) three faces
   (5) no flat surfaces

Use this figure to answer questions 2 and 3.

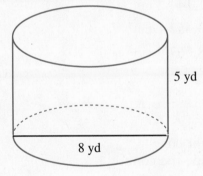

5 yd

8 yd

2. What is the total surface area?
   (1) 125.6 yd^2
   (2) 157 yd^2
   (3) 282.6 yd^2
   (4) 226.08 yd^2
   (5) 219.8 yd^2

3. What is the volume?
   (1) 251.2 yd^3
   (2) 1004.8 yd^3
   (3) 282.6 yd^3
   (4) 40 yd^3
   (5) 83.73 yd^3

Use this figure to answer questions 4 and 5.

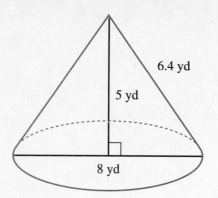

4. What is the total surface area?
   (1) 80.384 yd²
   (2) 130.624 yd²
   (3) 50.24 yd²
   (4) 200.96 yd²
   (5) 125.6 yd²

5. What is the volume?
   (1) 251.2 yd³
   (2) 753.6 yd³
   (3) 128 yd³
   (4) 1,004.8 yd³
   (5) 83.73 yd³

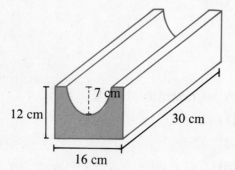

6. What is the volume of this figure?
   (1) 5,760 cm³
   (2) 4,615.8 cm³
   (3) 1,144.2 cm³
   (4) 3,452.1 cm³
   (5) 8,067.9 cm³

7. The figure below shows a right circular cylinder with a right cone inside it. What is the volume contained in the space outside the cone and inside the cylinder?

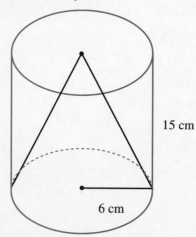

   (1) 565.2 cm³
   (2) 1,695.6 cm³
   (3) 2,260.8 cm³
   (4) 376.8 cm³
   (5) 1,130.4 cm³

10

Prisms and Pyramids

Circle the correct answer for each question.

1. The total surface area of a prism or a pyramid is _____ .
   - **(1)** the same as the lateral surface area
   - **(2)** the sum of the lateral surface area and the area of the base(s)
   - **(3)** the difference of the lateral surface area and the area of the base(s)
   - **(4)** the product of the areas of the surfaces
   - **(5)** equal to its volume

2. If a prism and a pyramid have the same base and the same height, the volume of the pyramid is _____ .
   - **(1)** equal to the volume of the prism
   - **(2)** three times the volume of the prism
   - **(3)** one-third the volume of the prism
   - **(4)** not related to the volume of the prism
   - **(5)** measured in square units

Use the figure below to answer questions 3–5.

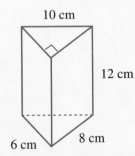

10 cm

12 cm

6 cm   8 cm

3. What is this figure called?
   - **(1)** rectangular solid
   - **(2)** triangular prism
   - **(3)** triangular pyramid
   - **(4)** regular pyramid
   - **(5)** rectangular prism

4. What is its surface area?

    (1) 96 cm²

    (2) 960 cm²

    (3) 288 cm²

    (4) 260 cm²

    (5) 336 cm²

5. What is its volume?

    (1) 576 cm³

    (2) 288 cm³

    (3) 5,760 cm³

    (4) 480 cm³

    (5) 1,440 cm³

Use the figure below to answer questions 6–8.

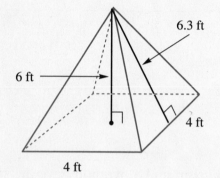

6.3 ft

6 ft

4 ft

4 ft

6. What is this figure called?

    (1) triangular pyramid

    (2) square pyramid

    (3) square prism

    (4) rectangular solid

    (5) equilateral pyramid

7. What is its total surface area?

    (1) 64 ft²

    (2) 96 ft²

    (3) 50.4 ft²

    (4) 604.8 ft²

    (5) 66.4 ft²

8. What is its volume?

    (1) 32 ft³

    (2) 96 ft³

    (3) 288 ft³

    (4) 24 ft³

    (5) 30.3 ft³

Parts of Prisms and Pyramids

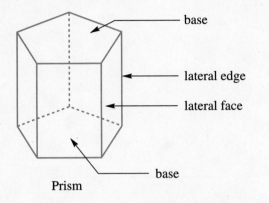

base

lateral edge

lateral face

base

Prism

A **prism** has two bases that are congruent polygons. The line segments joining the vertices of the bases form **lateral faces.** Adjacent lateral faces intersect in a straight line to form a lateral edge. A line segment perpendicular to both bases is called an **altitude.** The length of an altitude is called the **height** of the prism.

A prism is named by the shape of its bases.
If the bases are triangles, the prism is a **triangular prism.**
If the bases are rectangles, the prism is a **rectangular prism.**
If the bases are pentagons, the prism is a **pentagonal prism.**

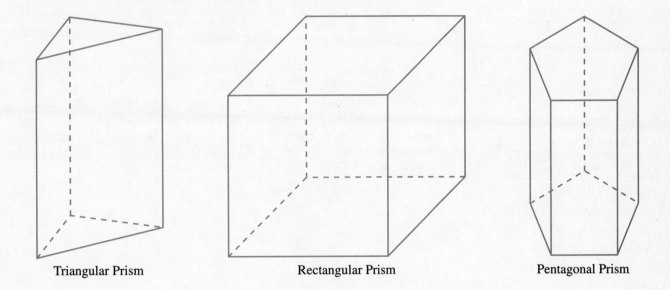

Triangular Prism   Rectangular Prism   Pentagonal Prism

If the lateral edges of a prism are perpendicular to its bases, the prism is a **right prism.** If the lateral edges are not perpendicular to the bases, the prism is an **oblique prism.**

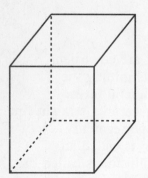

A rectangular prism has four lateral faces and two bases. Each of the lateral faces is a rectangle.

You can make models of prisms by cutting and folding paper. This pattern shows how to make a model of a rectangular prism. To make the model, cut along the solid lines and fold along the dotted lines.

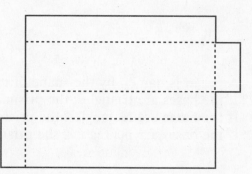

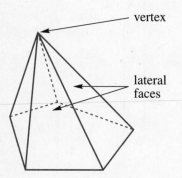

This is a pyramid. Its base is a polygon. The lateral faces are always triangles. The lateral faces intersect in a **vertex.** The line from the vertex perpendicular to the base is called the **altitude** of the pyramid. The length of the altitude is called the **height** of the pyramid.

Like a prism, a pyramid is named by the shape of its base.
If the base is a triangle, the pyramid is a **triangular pyramid.**
If the base is a rectangle, the pyramid is a **rectangular pyramid.**
If the base is a pentagon, the pyramid is a **pentagonal pyramid.**

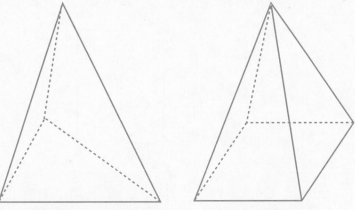 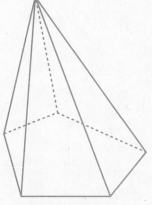

Triangular Pyramid   Rectangular Pyramid   Pentagonal Pyramid

If the base of a pyramid is a regular polygon and all of the
lateral edges are congruent, the pyramid is a **regular pyramid.**

Pyramids were built about 4,500 years ago as tombs for
Egyptian kings. These pyramids are large structures with square
bases. Since the base is a regular polygon, the Egyptian
pyramids are regular pyramids.

Use the figure below to answer questions 1–4.

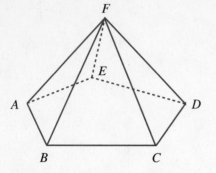

Use the figure below to answer questions 5–8.

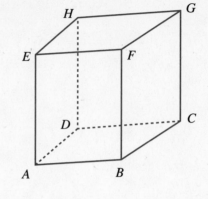

1. Name the vertex of the figure.

   _____

2. Name two lateral faces.

   _____

   _____

3. Identify the shape of the base.

   _____

4. Name the figure.

   _____

5. Name two faces.

   _____

   _____

6. Identify the shape of the bases.

   _____

7. Name three lateral edges.

   _____

   _____

   _____

8. Name the figure.

   _____

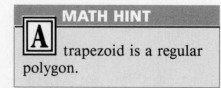

MATH HINT

A trapezoid is a regular polygon.

Surface Areas of Prisms and Pyramids

The **surface area of a prism** is the sum of the areas of the two bases and the areas of the lateral faces. The sum of the areas of the lateral faces is called the **lateral area.**

The lateral area of a right prism is given by the formula

$$LA = ph$$

where LA is the lateral area, p is the perimeter of the base, and h is the height of the prism.

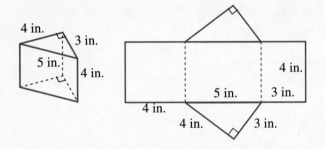

MATH HINT

R emember how the lateral area of a cylinder was found by laying flat the label portion of a can? The same reasoning applies here.

The formula for finding the surface area of a right prism is

$$SA = ph + 2B$$

where B is the area of base.

Example

A. Find the surface area of this figure.

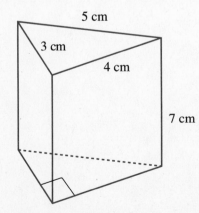

Find the lateral area first.

$$LA = ph$$
$$LA = (3 + 4 + 5)(7)$$
$$LA = 84$$

The lateral area of the prism is 84 cm².

Find the total area of the triangular bases. The formula for finding the area of a triangle is $A = \frac{1}{2} bh$. The total area of the two bases is 12.

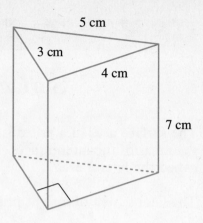

$$2B = 2\left(\tfrac{1}{2}\right) bh$$
$$2B = 2\left(\tfrac{1}{2}\right)(3)(4)$$
$$2B = 12$$

The total area of the triangular bases is 12 cm².

$$SA = LA + 2B$$
$$SA = 84 + 12$$
$$SA = 96$$

The surface area of the figure is 96 cm².

The **surface area of a pyramid** is the sum of the area of the base and the areas of the lateral faces. The sum of the areas of the lateral faces is called the **lateral area.**

The formula for finding the lateral area of a regular pyramid is

$$LA = \tfrac{1}{2}pl$$

where LA is lateral area, p is the perimeter of a base, and l is the height of a lateral face.

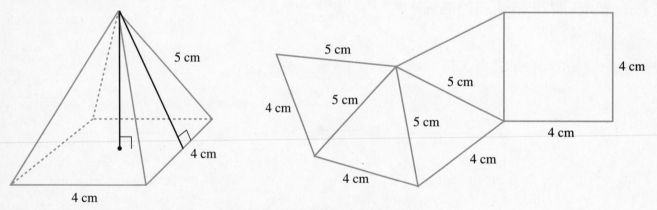

The formula for finding the surface area of a regular pyramid is:

$$SA = \tfrac{1}{2}pl + B$$

where B is the area of the base.

B. Find the surface area of this figure.

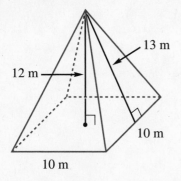

Find the lateral area first.

$$LA = \tfrac{1}{2}pl$$
$$LA = \tfrac{1}{2}(10 + 10 + 10 + 10)(13)$$
$$LA = \tfrac{1}{2}(40)(13)$$
$$LA = 260$$

The lateral area is 260 m².

Find the area of the square base. To find the area of the square, use the formula $B = lw$.

$$B = (10)(10)$$
$$B = 100$$

The area of the square base is 100 m².

$$SA = LA + B$$
$$SA = 260 + 100$$
$$SA = 360$$

The surface area of the figure is 360 m².

Practice

Use the figure below to answer questions 1–5.

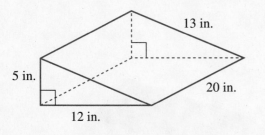

1. What kind of triangle is each base? _____

2. What is the height of each base? _____

3. Find the area of the bases. _____

4. Find the lateral area. _____

5. Find the total surface area. _____

Find the total surface area for problems 6–9.

6.

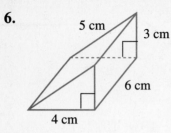

5 cm   3 cm
6 cm
4 cm

_____

7.

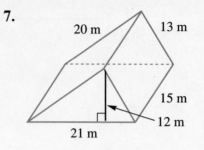

20 m   13 m
15 m
12 m
21 m

_____

8. The base is a square with sides measuring 12 inches.

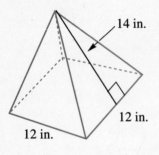

14 in.
12 in.
12 in.

_____

9. The base is an equilateral triangle with sides measuring 6 cm.

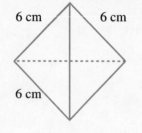

6 cm   6 cm
6 cm

_____

Problem Solving

Solve.

10. A painter needs to find the surface area to estimate the cost of painting a house. What is the surface area of the house?

_____

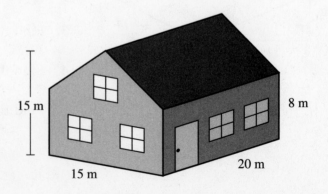

15 m
15 m
8 m
20 m

Volumes of Prisms and Pyramids

The volume of a prism is equal to the area of the base times the height. Use this formula

$$V = Bh$$

where B is the area of the base and h is the height.

To find the volume of a pyramid, use the formula

$$V = \frac{1}{3} Bh$$

where V is the volume,
B is the area of the base,
and h is the height.

Examples

A. Find the volume of this triangular prism. The base is a right triangle.

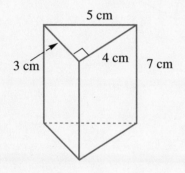

To find the area of the base, use the formula

$A = \frac{1}{2}bh$
$A = \frac{1}{2}(3)(4)$
$A = 6$

The area of the base is 6 cm².

$V = Bh$
$V = 6(7)$
$V = 42$

The volume of the figure is 42 cm³.

B. Find the volume of this regular square pyramid.

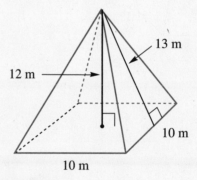

To find the area of the base, use the formula

$A = s^2$
$A = (10)(10)$
$A = 100$

The area of the base is 100 m².

$V = \frac{1}{3}(100)(12)$
$V = 400$

The volume of the pyramid is 400 m³.

Find the volumes of the following figures.

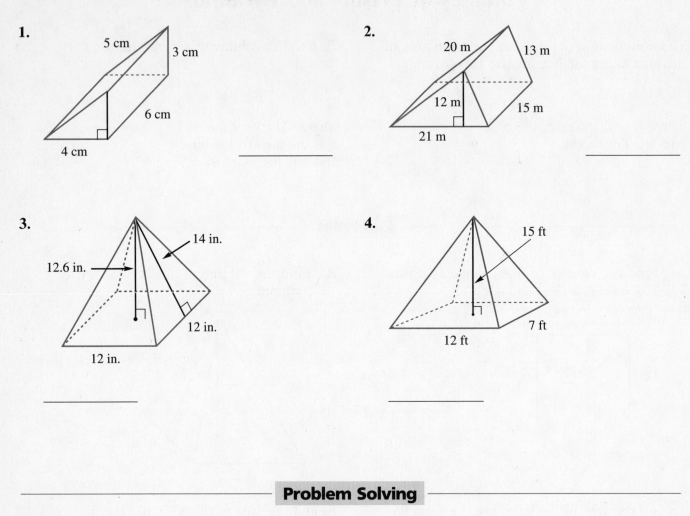

1.

5 cm
3 cm
6 cm
4 cm

_____

2.

20 m
13 m
12 m
15 m
21 m

_____

3.

14 in.
12.6 in.
12 in.
12 in.

_____

4.

15 ft
7 ft
12 ft

_____

Problem Solving

Solve.

5. The ends of a metal bar are parallel isosceles trapezoids. Find the volume of the bar.

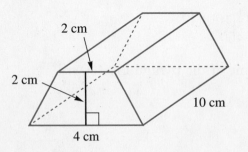

2 cm
2 cm
10 cm
4 cm

_____

6. The pyramid of Khufu, or the Great Pyramid, was originally 481 feet tall with a square base that measured 755 feet on each side. What was the pyramid's volume?

_____

Problem Solving: Drawing a Picture

The steps you have learned to help solve word problems can be used with word problems that deal with geometric figures. Use the following steps:

Step 1 Read the problem and underline the key words. These words will generally relate to some mathematical reasoning.

Step 2 Make a plan to solve the problem. Ask yourself, Should I add, subtract, multiply, divide, round, or compare? You may have to do more than one of these operations for the same problem.

Step 3 Find the solution. Use your math knowledge to find your answer.

Step 4 Check your answer. Ask yourself, Is the answer reasonable? Did you find what you were asked for?

When working with geometric figures, it is always helpful to draw the figure and label the figure with the information given in the problem. Seeing the figure will help you better decide what you need to find and what you have.

Example

A swimming pool has a shallow end and a deep end. The shallow end starts at a depth of 3 feet and tapers to a depth of 9 feet. The deep end stays at 9 feet. The pool is 18 feet wide. The deep end is 12 feet long. The part of the pool that tapers is 16 feet long. What is the volume of pool?

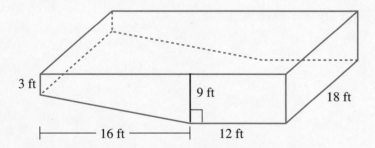

Step 1 The key word is **volume.**

Step 2 First draw the pool to help you see what is given.

Drawing the pool helps you see that the pool can be divided into two figures: a rectangular prism and a trapezoidal prism. Find the areas of both prisms.

Step 3 To find the volume of the rectangular prism, use the following formula

$$V = lwh$$
$$V = 18(12)(9)$$
$$V = 1,944$$

The volume of the rectangular prism is 1,944 cubic feet.

To find the volume of the trapezoidal prism, use this formula:

$$V = Bh$$

The area of the base is:

$$B = \tfrac{1}{2}(3 + 9)(16)$$
$$B = 96$$

The area of the base is 96 square feet.

> **MATH HINT**
>
> **T**he formula for the area of a trapezoid is $A = \tfrac{1}{2}(b_1 + b_2)h$.

$$V = 96(18)$$
$$V = 1,728$$

The volume of the trapezoidal prism is 1,728 cubic feet.

The total volume of the pool is (1,944 + 1,728), or 3,672 cubic feet.

Step 4 Check your answer.

If the pool were a rectangular prism 18 feet wide by 28 feet long and 9 feet deep, the volume would be 4,536 cubic feet.

If the pool were a rectangular prism 18 feet wide by 28 feet long and 3 feet deep, the volume would be 1,512 cubic feet.

The answer is between these numbers. The answer is reasonable.

1. The top of a house which is 40 ft × 60 ft is in the shape of an equilateral triangular prism. If one package of shingles covers 100 square feet, how many packages of shingles do you need to cover the roof of the house? (Hint: Draw a figure before you solve the problem.) _____

2. A paving block is in the shape of a regular hexagonal prism. Each side measures 8 inches and the height is 2 inches. How much concrete is needed to make each block? (Hint: Draw the block before you solve the problem.) _____

Cost of Filling a Swimming Pool

Myra has a rectangular swimming pool that is 10 feet wide by
30 feet long, with a constant depth of 4 feet. She needs to fill the
pool with water. If her water company charges $2.00 per 100 cubic
feet of water, how much will she pay to fill her pool?

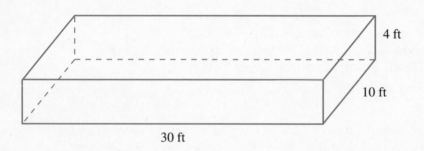

To find the volume of the pool, use the formula $V = lwh$.

$V = 10(30)(4)$
$V = 1,200$

The volume of the pool is 1,200 cubic feet of water.

The cost of the water is $2 per 100 cubic feet. To find the total cost
of the water, divide the total cubic feet by 100. Then multiply that
answer by $2.

Cost $= (1200 \div 100) \times \$2$
Cost $= 12 \times \$2$
Cost $= \$24.00$

The total cost to fill the swimming pool is $24.00.

Answer each of the following questions.

1. Sometimes companies figure out your cost of water per gallon. If one cubic foot of water equals about 7.5 gallons, how many gallons of water does Myra's pool hold?

2. Lee's bathtub is 24 inches wide, 4.5 feet long, and 12 inches deep. What is the volume of the tub in cubic feet and in gallons? (Hint: Change inch measurements to feet.)

_____

_____

Posttest

Circle the best answer for each question.

1. A pyramid has _____.
   (1) two regular polygons as bases
   (2) a vertex and one regular polygon as a base
   (3) a vertex and a polygon as a base
   (4) no more than three faces
   (5) two congruent polygons as bases

2. The lateral area of a prism or a pyramid _____.
   (1) is the same as the surface area
   (2) does not include the area of the base(s)
   (3) does include the area of the base(s)
   (4) is the product of the areas of the surfaces
   (5) is equal to its volume

3. If a prism and a pyramid have the same base and the same height, the volume of the prism is _____.
   (1) equal to the volume of the pyramid.
   (2) three times the volume of the pyramid.
   (3) one-third the volume of the pyramid.
   (4) not related to the volume of the pyramid.
   (5) measured in square units.

Use the figure below to answer questions 4–6.

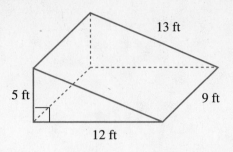

13 ft

5 ft   9 ft

12 ft

4. What is this figure called?

   (1) rectangular solid
   (2) triangular prism
   (3) triangular pyramid
   (4) regular pyramid
   (5) rectangular prism

5. What is its surface area?

   (1) 330 ft²
   (2) 300 ft²
   (3) 39 ft²
   (4) 540 ft²
   (5) 177 ft²

6. What is its volume?

   (1) 7,020 ft³
   (2) 39 ft³
   (3) 540 ft³
   (4) 270 ft³
   (5) 180 ft³

Use the figure below to answer questions 7–9.

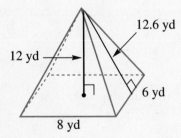

12.6 yd

12 yd

6 yd

8 yd

7. What is this figure called?

   (1) triangular pyramid
   (2) square pyramid
   (3) square prism
   (4) rectangular solid
   (5) rectangular pyramid

8. What is its total surface area?

   (1) 576 yd²
   (2) 168 yd²
   (3) 216 yd²
   (4) 26 yd²
   (5) 224.4 yd²

9. What is its volume?

   (1) 192 yd³
   (2) 583.76 yd³
   (3) 51.02 yd³
   (4) 576 yd³
   (5) 1,728 yd³

UNIT

11

Spheres

Circle the best answer for each question.

1. The radius of a sphere is _____ .

    (1) two times the size of the diameter

    (2) one-half of the diameter

    (3) the distance across the largest circle

    (4) larger than the radius of the largest circle

    (5) measured in cubic units

2. What is half of a sphere called?

    (1) hemisphere

    (2) semisphere

    (3) spheroid

    (4) circle

    (5) radius

3. The diameter of a sphere is 11 cm. What does its radius measure?

    (1) 22 cm

    (2) 6.5 cm

    (3) 5.5 cm

    (4) 4.5 cm

    (5) 3.6 cm

4. What is the surface area of a sphere that has a diameter of 24 cm?

    (1) 7,234.56 cm^2

    (2) 1,808.64 cm^2

    (3) 602.88 cm^2

    (4) 2,411.52 cm^2

    (5) 12 cm

5. The surface area of a hemisphere
is _____ .

    (1) two times the surface area of a sphere
with the same radius

    (2) four times the surface area of a sphere
with the same radius

    (3) one-half the surface area of a sphere
with the same radius

    (4) four-thirds the surface area of a sphere
with the same radius

    (5) equal to the volume

6. What is the surface area of a sphere that
has a radius of 12 cm?

    (1) 904.32 cm^2

    (2) 1,808.64 cm^2

    (3) 3,617.28 cm^2

    (4) 1,205.76 cm^2

    (5) 7,234.56 cm^2

7. What is the volume of a sphere that has a
diameter of 24 cm?

    (1) 2,411.52 cm^3

    (2) 768 cm^3

    (3) 21,703.68 cm^3

    (4) 57,876.48 cm^3

    (5) 7,234.56 cm^3

8. What is the volume of a sphere that has a
radius of 12 cm?

    (1) 7,234.56 cm^3

    (2) 28,938.24 cm^3

    (3) 10,851.84 cm^3

    (4) 14,469.12 cm^3

    (5) 1,205.76 cm^3

9. What is the volume of this figure?

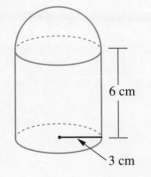

6 cm

3 cm

    (1) 18 cm^3

    (2) 113.04 cm^3

    (3) 169.56 cm^3

    (4) 226.08 cm^3

    (5) 339.12 cm^3

Parts of Spheres

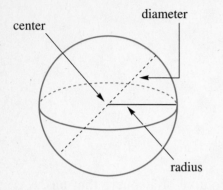

center
diameter
radius

A **sphere** is a solid with all points the same distance from the center. A line segment joining the center and any point on the sphere is a **radius**. A **diameter** of the sphere is a line segment passing through the center with endpoints on the sphere.

A **hemisphere** is one-half of a sphere. The radius and the diameter of a hemisphere are the same measures as the radius and diameter of the sphere itself.

Examples of spheres are a baseball, a globe, and a tennis ball.

Example

Find the radius of a sphere with a diameter of 18 mm.

Use the formula $r = \frac{1}{2}d$.
$r = \frac{1}{2}(18)$
$r = 9$
The radius of the sphere is 9 mm.

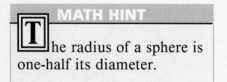

MATH HINT

The radius of a sphere is one-half its diameter.

Use the figure below to answer questions 1–4.

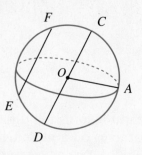

1. Name a radius of the sphere.

    _____

2. Name a diameter of the sphere.

    _____

3. If the radius measures 10 cm, how long is the diameter?

    _____

4. If the diameter measures 12 m long, how long is the radius?

    _____

Problem Solving

Solve.

5. If the diameter of a sphere is 4 cm, what is the diameter of one-half the sphere, or the hemisphere?

    _____

6. The radius of a hemisphere is 14 inches. What is the diameter of the hemisphere?

    _____

LIFE SKILL

Map Lines

Lines are drawn on maps of the earth to help us tell directions and to locate places. There are two sets of lines on a map. One set is called lines of **latitude.** Lines of latitude run east and west. Each line of latitude has a number that is measured in degrees (°). The **equator** is a line of latitude that circles the earth midway between the North Pole and the South Pole. The equator is 0° latitude. The lines of latitude can be used to measure distances north or south of the equator. As you can see on the map, lines of latitude never meet; therefore, they are **parallels.**

The second set of lines on a map, which run north and south, are called lines of **longitude.** Lines of longitude can be used to measure distances east or west from the starting line, numbered 0° longitude. This starting line is called the **prime meridian.** This line passes through Greenwich, England.

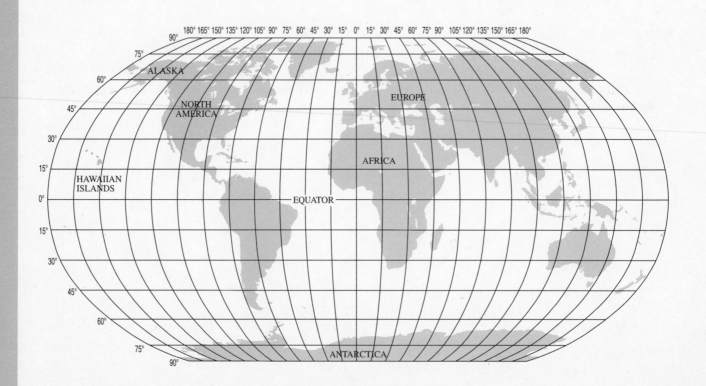

Answer each of the following questions.

1. The United States is located between which two lines of latitude?

   _____

2. Alaska is located between which two lines of latitude?

   _____

3. The Hawaiian Islands are between which degrees of latitude?

   _____

4. Antarctica is between which degrees of latitude?

   _____

5. What continents does the prime meridian go through?

   _____

6. Between which two lines of longitudes is the state of Florida?

   _____

7. The diameter of the earth at the equator is 7,926.41 miles. What is the circumference of the earth at the equator?

   _____

8. The diameter of the earth at the prime meridian is 7,899.83 miles. What is the circumference of the earth at this location?

   _____

Surface Areas of Spheres

To find the surface area of a sphere, you can use the formula

$$SA = 4\pi r^2$$

where SA is the surface area and r is the radius.

Examples

A. Find the surface area of a sphere with a diameter of 12 m.

$$SA = 4\pi r^2$$
$$SA = 4\pi(6)^2$$
$$SA = 4(3.14)(36)$$
$$SA = 452.16$$

MATH HINT
T he radius of a sphere is one-half the diameter.

The surface area of the sphere is 452.16 square meters.

B. The surface area of a sphere is 200.96 square meters. What is the radius of the sphere?

$$SA = 4\pi r^2$$
$$200.96 = 4(3.14)r^2$$
$$200.96 = 12.56r^2$$
$$16 = r^2$$
$$4 = r$$

The radius of the sphere is 4 meters.

Practice

For problems 1–4, use the information given to find the surface area of
each sphere.

1. radius = 10 mm _____

2. radius = 5 ft _____

3. diameter = 16 m _____

4. diameter = 42 ft _____

Solve.

5. Find the surface area of a sphere whose radius is 21 ft. _____

6. The diameter of the earth at the equator is 7,926.41 miles. Find the surface area of the earth. _____

7. The surface area of a sphere is 11,304 in.^{2} Find the radius of the sphere. _____

Use this information to answer questions 8–10:
A sphere has a radius of 7 mm.

8. Find the ratio of the surface areas of this sphere and another sphere whose radius is double that of the original sphere. _____

9. Find the ratio of the surface areas of this sphere and another sphere whose radius is triple that of the original sphere. _____

10. Find the ratio of the surface areas of this sphere and another sphere whose radius is one-half that of the original sphere. _____

Volumes of Spheres

To find the volume of a sphere, use the following formula:

$$V = \frac{4}{3} \pi r^3$$

where V is the volume and r is the radius of the sphere.

Examples

A. Find the volume of a sphere with a diameter of 12 meters.

$$V = \frac{4}{3} \pi r^3$$
$$V = \frac{4}{3}(3.14)(6)^3$$
$$V = 904.32$$

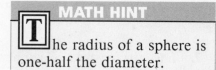

MATH HINT

The radius of a sphere is one-half the diameter.

The volume of the sphere is 904.32 cubic meters.

B. The volume of a sphere is 268.08 cubic meters. Find the radius of the sphere.

$$V = \frac{4}{3} \pi r^3$$
$$268.08 = \frac{4}{3}(3.14)r^3$$
$$268.08 = 4.187r^3$$
$$64.027 = r^3$$
$$4 = r$$

The radius of the sphere is about 4 meters.

Practice

For problems 1–4, use the information given to find the volume of each sphere. Round answers to the nearest hundredth.

1. radius = 10 mm _____

2. radius = 5 ft _____

3. diameter = 16 m _____

4. diameter = 42 ft _____

Solve.

5. Find the volume of a sphere whose radius is 21 ft. _____

6. The diameter of the earth at the equator is 7,926.41 miles. Find the volume of the earth. _____

7. The volume of a sphere is 113,040 in.^{3} Find the radius of the sphere. _____

Use this information to answer questions 8–10:
A sphere has a radius of 7 mm.

8. Find the ratio of the volumes of this sphere and another sphere whose radius is double that of the original sphere. _____

9. Find the ratio of the volumes of this sphere and another sphere whose radius is triple that of the original sphere. _____

10. Find the ratio of the volumes of this sphere and another sphere whose radius is one-half that of the original sphere. _____

Problem Solving: Hidden Facts

The steps you have learned to help solve word problems can be used with word problems that deal with geometric figures. Use the following steps:

Step 1   Read the problem and underline the key words. These words will generally relate to some mathematical reasoning.

Step 2   Make a plan to solve the problem. Ask yourself, Should I add, subtract, multiply, divide, round, or compare? You may have to do more than one of these operations for the same problem.

Step 3   Find the solution. Use your math knowledge to find your answer.

Step 4   Check your answer. Ask yourself, Is the answer reasonable? Did you find what you were asked for?

Sometimes there are hidden facts in a problem. For example, sometimes a problem gives the diameter to find the volume of a sphere. The radius is easily found by dividing by 2. This is a hidden fact since it is implied and not actually given to you.

Example

A silo is shaped like a cylinder with half a sphere placed on top. The diameter of the silo is 14 feet. The height of the cylinder is 20 feet. Find the volume of the silo.

Step 1   The key words are **volume, diameter,** and **height.**

Step 2   Find the volume of each part of the silo, and add the volumes together.

Hidden facts: The radius of the cylinder and the sphere is $14 \div 2 = 7$.

Step 3   To find the volume of the cylinder, do the following:

$$V = \pi r^2 h$$
$$V = \tfrac{22}{7} \times 7 \times 7 \times 20$$
$$V = 3{,}080$$

The volume of the cylinder is about 3,080 cubic feet.

To find the volume of the part of the silo that is half a sphere, do the following:

$$V = \frac{1}{2}\left(\frac{4}{3}\right)\pi r^3$$
$$V = \frac{2}{3} \times \frac{22}{7} \times 7 \times 7 \times 7$$
$$V = 718\frac{2}{3}$$

The volume of the top is about 718 cubic feet.

To find the total volume of the silo, add the volumes of the cylinder and the half sphere:

$$3,080 + 718\frac{2}{3} = 3,798\frac{2}{3}$$

The total volume of the silo is about $3,798\frac{2}{3}$ cubic feet.

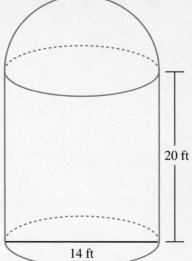

20 ft

14 ft

Step 4   Check to see if the answer is reasonable.

If the silo were a cylinder with a total height of 27 feet (20 + 7), the volume would be 4,158 cubic feet.

$$V = \frac{22}{7} \times 7 \times 7 \times 27$$
$$V = 4158 \text{ cu ft}$$

If the silo were a cylinder 20 feet tall and had no half sphere on top, the volume would be 3,080 cubic feet. The answer is between 4,158 and 3,080.

Practice

Use the figure below to answer questions 1–2.

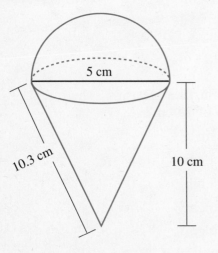

5 cm

10.3 cm

10 cm

1. Find the surface area of the figure.

_____

2. Find the total volume of the figure.

_____

Posttest

Circle the best answer for each question.

1. The diameter of a sphere is _____.
   (1) two times the size of the radius
   (2) one-half of the radius
   (3) the distance from the center of a sphere to the outside
   (4) smaller than the diameter of a great circle
   (5) measured in cubic units

2. A sphere is a solid _____.
   (1) with a diameter four times the radius.
   (2) perpendicular to a plane.
   (3) with a radius that is four times the diameter.
   (4) that looks like an ice cone.
   (5) with all points the same distance from the center.

3. The radius of a sphere is $3\frac{1}{2}$ in. What is its diameter?
   (1) $1\frac{3}{4}$ in.
   (2) 14 in.
   (3) 7 in.
   (4) $5\frac{1}{2}$ in.
   (5) $6\frac{1}{2}$ in.

4. What is the surface area of a sphere that has a diameter of 48 cm?
   (1) 7,234.56 cm²
   (2) 1,808.64 cm²
   (3) 602.88 cm²
   (4) 2,411.52 cm²
   (5) 12 cm

5. When the radius of a sphere is doubled, the surface area _____.
   (1) doubles
   (2) remains the same
   (3) is squared
   (4) is four times as great
   (5) is equal to the volume

6. What is the surface area of a sphere that has a radius of 24 cm?
   (1) 904.32 cm²
   (2) 1,808.64 cm²
   (3) 3,617.28 cm²
   (4) 1,205.76 cm²
   (5) 7,234.56 cm²

7. What is the volume of a sphere that has a diameter of 48 cm?

    (1) 2,411.52 cm³

    (2) 768 cm³

    (3) 21,703.68 cm³

    (4) 57,876.48 cm³

    (5) 7,234.56 cm³

8. What is the volume of a sphere that has a radius of 10 cm?

    (1) 3,617.28 cm³

    (2) 28,938.24 cm³

    (3) 10,851.84 cm³

    (4) 14,469.12 cm³

    (5) 4,186.67 cm³

9. When the radius of a sphere is doubled, the volume _____ .

    (1) doubles

    (2) remains the same

    (3) is eight times as great

    (4) is four times as great

    (5) is equal to the surface area

Unit 1 Pretest/pages 1-2

1. (2)
2. (4)
3. (1)
4. (3)
5. (1)
6. (3)
7. (5)
8. (2)
9. (4)
10. (3)

Lesson 1/pages 3-6

1. 179°, no
2. 170°, no
3. 180°, yes
4. 180°, yes
5. 190°, no
6. 149°, no
7. 85°
8. 281,250 ft^2
9. 112 ft 2 in.
10. 72 in.
11. 2,150 ft

Lesson 2/pages 7-13

1. $8\frac{1}{2}$ ft
2. $5\frac{1}{3}$ ft
3. $13\frac{5}{6}$ ft
4. $7\frac{1}{3}$ in.^{2}
5. $4\frac{3}{8}$ in.
6. $41\frac{17}{24}$ in.
7. $2\frac{3}{4}$ in.
8. $3\frac{13}{16}$ in.
9. 19 floorboards
10. $156\frac{1}{4}$ pounds

Lesson 3/pages 14-18

1. 11.75 mm
2. 4.2 cm
3. 91.88 ft^2
4. 3 pieces, 0.5 ft
5. 10.4 m

Lesson 4/pages 19-22

1. $4\frac{2}{3}$ ft
2. 30 in.
3. 7.9 mm

4. $56\frac{1}{4}$ miles
5. 140 miles
6. 36 teeth
7. 80.6 km
8. 9 in.
9. $1\frac{1}{3}$ gallons
10. 5 gallons

Lesson 5/pages 23-27

1. 1
2. 1
3. 32
4. 1
5. 15
6. 1.7321
7. 2.4495
8. 27
9. 1,296
10. 225
11. 706.5 mm^2
12. 2.62 m
13. 10 in.
14. 5.29 mm
15. $a^2 + b^2 = c^2$
   $a^2 + 12^2 = 15^2$
   $a^2 = 225 - 144$
   $a^2 = 81$
   $a = \sqrt{81} = 9$
16. $a^2 + b^2 = c^2$
   $(20.25)^2 + (36)^2 = c^2$
   $410.0625 + 1,296 = c^2$
   $1706.0625 = c^2$
   $41.30 = c$

Lesson 6/pages 28-29

1. $2.40
2. $60
3. $116.76
4. $95.88
5. $6.56
6. 12 sq yd, $115.08

Unit 1 Posttest/pages 30-31

1. (3)
2. (2)
3. (1)
4. (4)
5. (3)
6. (5)
7. (3)
8. (2)
9. (5)
10. (1)

Unit 2 Pretest/pages 32-33

1. 18°
2. An obtuse angle
3. They are equal.
4. Right angle
5. (2)
6. (4)
7. (2)
8. 180°
9. (5)
10. (3)
11. (1)
12. (1)

Lesson 7/pages 34-37

1. $\vec{EA}$, $\vec{ED}$, $\vec{EC}$, $\vec{EB}$
2. E
3. $\angle AED$, $\angle DEC$, $\angle CEB$, $\angle BEA$
4. $\overline{AE}$, $\overline{EC}$, $\overline{DE}$, $\overline{EB}$, $\overline{BD}$, $\overline{AC}$
5. $\overleftrightarrow{AC}$, $\overleftrightarrow{BD}$
6. They are intersecting lines because they meet in a point.
7. No. $\vec{AE}$ has A as an endpoint. $\vec{EA}$ has E as an endpoint.
8. Yes. E is the vertex of the angle with sides EA and ED.
9. 1
10. No

Lesson 8/pages 38-41

1. 90°
2. 110°
3. 50°
4. 85°

Use your protractor to compare these angles to those you have drawn.

5.

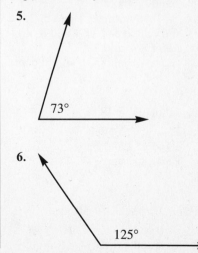

73°

6.

125°

7.

8.

9. Obtuse
10. Acute
11. Right
12. Obtuse

Lesson 9/pages 42-44

1. 55°
2. 55°
3. Vertical angles
   They are equal.
4. Vertical angles
   They are equal.
5. $x = 128°$
6. $x = 20°$
7. $x = 4°$
8. $x = 75°$

Life Skill/page 45

1. 45°, left
2. 60°, left
3. 15°, left
4. 22°, right

Lesson 10/pages 46-47

1. ∠s 3 and 7 equal 141°. ∠8 = 39°
2. Any two of these: 3, 6, or 7
3. 3 and 6, 4 and 5
4. Any two of these: 1 and 5, 2 and 6, 3 and 7, or 4 and 8
5. 1 and 8, 2 and 7
6. 1 and 4, 2 and 3, 5 and 8, 6 and 7

Lesson 11/page 48

1. They are parallel.
2. 3 and 6, 4 and 5
3. 1 and 8, 2 and 7
4. 90°

Lesson 12/pages 49-50

1. ∠s a and c = 50°, ∠b = 130°
2. ∠s a and c = 120°, ∠b = 60°

Unit 2 Posttest/pages 51-52

1. (1)
2. (3)
3. (3)
4. (2)
5. (2)
6. (4)
7. (4)
8. (3)
9. (3)
10. (3)

Unit 3 Pretest/pages 53-54

1. (2)
2. (3)
3. (2)
4. (1)
5. (2)
6. (2)
7. (2)
8. (5)
9. (3)
10. (1)

Lesson 13/pages 55-59

1. Isosceles, 90°
2. Scalene, 115°
3. Scalene, 20°
4. Scalene, 88°
5. Isosceles, 130°
6. Scalene, 112°
7. 60°
8. Both are 55°.
9. 60°
10. 90°

Lesson 14/pages 60-62

1. 15 in.
2. 15 yd 4 ft = 16 yd 1 ft
3. 7m 15 cm or 7.15 m
4. 14 ft
5. 2.8 m
6. 3 ft 2 in.
7. 34 ft
8. 8 in.
9. True
10. Shortest

Lesson 15/pages 63-64

1. 56 sq in.
2. 7.5 or $7\frac{1}{2}$ sq ft
3. 22.5 or $22\frac{1}{2}$ sq mm
4. 48 sq yd
5. 36 sq ft

Lesson 16/pages 65-66

1. 35°
2. 10
3. False. Similar triangles may be congruent. Congruent triangles are similar.
4. $\overline{DE}$
5. (3)
6. (4)
7. (1)
8. (2)

Lesson 17/pages 67-71

1. $\frac{RQ}{RT} = \frac{PQ}{ST} = \frac{RP}{RS}$
2. $\frac{MN}{PQ} = \frac{NO}{QO} = \frac{OM}{OP}$
3. Martha's is correct.
4. Martha: $FG = 14$;
   Jesse: $FG = 2\frac{4}{7}$.
5. 3 cm
6. 4.5 m
7. 12
8. 40 ft
9. 71.5 or $71\frac{1}{2}$ ft
10. 24 ft

Lesson 18/pages 72-74

1. no
2. yes
3. no
4. no
5. (1)
6. (2)
7. (3)
8. (3)
9. (1)
10. (3)

Life Skill/page 75

1. 10 ft
2. 34 in.

Lesson 19/pages 76-77

1. 1.2 quarts
2. 15 ft
3. 60 in.

Life Skill/pages 78-79

1. 22 ft, more efficient
2. 24 ft

Unit 3 Posttest/pages 80-81

1. (5)
2. (3)
3. (2)
4. (3)

5. (1)
6. (1)
7. (2)
8. (5)
9. (1)
10. (2)

Unit 4 Pretest/pages 82-83

1. (5)
2. (1)
3. (4)
4. (2)
5. (3)
6. (3)
7. (2)
8. (5)
9. (3)
10. (1)

Lesson 20/pages 84-87

1. (1) 60°
   (2) 180°
   (3) 120°
   (4) 360°
2. (1) 90°
   (2) 360°
   (3) 90°
   (4) 360°
3. (1) 108°
   (2) 540°
   (3) 72°
   (4) 360°
4. (1) 120°
   (2) 720°
   (3) 60°
   (4) 360°
5. (1) 128.6°
   (2) 900°
   (3) 51.4°
   (4) 360°
6. (1) 135°
   (2) 1080°
   (3) 45°
   (4) 360°
7. (1) The size of the interior angles increases as the number of sides increases.
   (2) The sum of the exterior angles of a regular polygon is 360°.
8. 15 sides

Life Skill/page 88

1. Quadrilateral
2. Octagon

Lesson 21/pages 89-90

1. 40 cm
2. 25 yd

3. 18 m
4. 10 ft
5. 80 in. or 6 ft 8 in.
6. 8.1 cm
7. 65 ft
8. 12.6 cm
9. 3.6 m
10. 1 m 51 cm or 1.51 m
11. Hexagon
12. The perimeter increases.

Lesson 22/pages 91-92

1. 2.61 in.^{2}
2. 3.64 m^2
3. 4.84 m^2
4. 7.70 m^2
5. 4,605 ft
6. 1,459,000 ft^2

Lesson 23/page 93

1. 20 in.
2. 389.7 m^2

Lesson 24/pages 94-95

1. 62 tiles (rounding to a full tile)
2. 35 bricks

Unit 4 Posttest/pages 96-97

1. (5)
2. (2)
3. (1)
4. (2)
5. (1)
6. (3)
7. (3)
8. (2)
9. (4)
10. (3)

Unit 5 Pretest/pages 98-99

1. (5)
2. (3)
3. (1)
4. (3)
5. (1)
6. (5)
7. (1)
8. (3)
9. (2)
10. (3)

Lesson 25/pages 100-103

1. $\angle C = 80°$, $\angle$s B and $D = 100°$
2. All four angles = 90°
3. $\angle G = 40°$, $\angle$s F and $H = 140°$
4. $\angle X = 60°$, $\angle$s Y and $Z = 120°$
5. 5 cm

Lesson 26/pages 104-106

1. (1) $10\frac{3}{4}$ in.
   (2) $8\frac{1}{2}$ in.
   (3) $38\frac{1}{2}$ in.
   (4) Rectangle
2. (1) Answer varies
   (2) Answer varies
   (3) Sum of 2(*height*) and 2(*width*)
   (4) Rectangle
3. 16 cm
4. 25 cm
5. 16 cm
6. 12 cm
7. 20 ft
8. 24 m
9. 460 m
10. 67 ft
11. 8 in.
12. 1,700 m

Lesson 27/pages 107-110

1. 21 cm^2
2. 9 ft^2 or 1 yd^2
3. 18 in.^{2}
4. 98,700 cm^2 or 9.87 m^2
5. 32 m^2
6. $6\frac{3}{4}$ in.^{2}
7. 96 sq ft
8. 96 sq ft
9. 650 trees
10. about 266 children

Lesson 28/pages 111-113

1. 16 sq ft
2. 56 ft
3. 236 ft
4. 8,220 sq ft
5. 2.8625 or 3 rolls

Lesson 29/pages 114-116

1. $10.92
2. 3 times
3. 100 lb.
4. 40.61 acres
5. 4 times

Life Skill/page 117

1. $7,875
2. $656.25
3. $55.25
4. $185.91
5. The factory building would be cheaper to rent, but cost more to move into.

Unit 5 Posttest/pages 118-119

1. (5)
2. (5)
3. (1)
4. (3)
5. (2)
6. (5)
7. (3)
8. (2)
9. (2)
10. (3)

Unit 6 Pretest/pages 120-121

1. (1)
2. (3)
3. (4)
4. (2)
5. (2)
6. (4)
7. 15.7 or $15\frac{5}{7}$ ft
8. 13.56 or $13\frac{4}{7}$ ft

Lesson 30/pages 122

1. Any two of these: $\overline{OC}$, $\overline{OD}$, $\overline{OA}$, or $\overline{OB}$
2. AB
3. AB and EF
4. any of these: ∠s COB, BOD, DOA, AOC (any angle with O as the vertex)
5. 10 inches
6. 15

Lesson 31/pages 123-125

1. 5, 31.4, 3.14
2. 8, 25.12, 3.14
3. 10, 20, 3.14
4. $17\frac{1}{2}$, 35, $3\frac{1}{7}$
5. 3.14 or $3\frac{1}{7}$
6. 7.85 in.
7. 9.42 in.
8. 10.99 ft
9. 681.38 or 682 ft
10. No, the diameter of the bolt is $\frac{3}{16}$ or 0.1875 in. but the diameter of the hole is only 0.093 in.
11. 163.28 or $163\frac{3}{7}$ in.

Life Skill/pages 126-127

1. $1,800,000
2. $450,000

Lesson 32/pages 128-130

1. 14 cm, 153.86 sq cm
2. 3 cm, 7.065 sq cm

3. 9 yd, 254.34 or $254\frac{4}{7}$ sq yd
4. 5 ft, 78.5 or $78\frac{4}{7}$ sq ft
5. 3.5 m, 38.465 sq m
6. $11\frac{1}{3}$ in., 100.83 sq in.
7. $4\frac{1}{12}$ mi, 52.36 sq mi
8. 2.25 km, 15.9 sq km
9. $\frac{1}{2}$ mi, 1 mi
10. 0.37 m, 0.74 m
11. 706.5 sq ft
12. 6358.5 sq km
13. 19.6 sq ft
14. 11,304 sq mi
15. 2.5 lb
16. 13 people

Lesson 33/pages 131-132

1. 27.42 in., 50.13 in.^{2}
2. 6.512 m, 1.0976 m^2
3. 27.42 in., 21.87 in.^{2}
4. 6.512 m, 2.1024 m^2

Lesson 34/pages 133-135

1. 16 ft
2. 200.96 ft^2
3. 178.98 ft^2
4. 0.059141 sq mi
5. 1.1425 mi
6. $2,892.50
7. 3.925 qt

Life Skill/pages 136-137

1. $14,875
2. $8,500
3. (1) $560.14
   (2) $641.16

Unit 6 Posttest/pages 138-139

1. (1)
2. (1)
3. (1)
4. (1)
5. (3)
6. (1)
7. $9\frac{13}{14}$ in.
8. 70 in.

Unit 7 Pretest/pages 140-141

1. (3)
2. (2)
3. (2)
4. (5)
5. (3)
6. (1)
7. (4)

8. (5)
9. (1)
10. (2)

Lesson 35/pages 142-144

1. (0,0)
2. (10,0)
3. (8,8)
4. (0,8)
5. (−6,5)
6. (−3,0)
7. (−4,−5)
8. (0,−9)
9. (8,−9)
10. (10,−2)

11-20.

Lesson 36/pages 145-147

1. 4
2. 7
3. 11
4. 5
5. About 8.5
6. About 3.6
7. 10
8. About 9.1
9. About 12
10. About 11.2

Lesson 37/pages 148-149

1. $(6, 3\frac{1}{2})$
2. $(-2\frac{1}{2}, 2\frac{1}{2})$
3. $(1, 3\frac{1}{2})$
4. $(-2\frac{1}{2}, -2)$
5. (5, 5)
6. $(7\frac{1}{2}, 0)$
7. (10, 6)
8. (−4, −2)

Lesson 38/pages 150-151

1. Perimeter: 24.2; area: 27

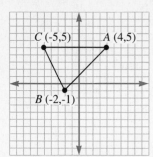

2. Perimeter: 60; area: 224

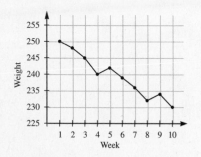

Lesson 39/pages 152-154

1. She will pay nothing because the $75 deposit covers the cost of the rental.
2. $760
3. about 20.8 miles
4. $110

Life Skill/page 155

Unit 7 Posttest/pages 156-157

1. (2)
2. (4)
3. (2)
4. (4)
5. (5)
6. (3)
7. (1)
8. (4)
9. (1)
10. (3)

Unit 8 Pretest/pages 158-159

1. (5)
2. (1)
3. (3)
4. (2)
5. (5)
6. (4)
7. (3)
8. (1)
9. (2)
10. (4)

Lesson 40/pages 160-162

1. Three pairs: *ABFE* and *DCGH*; *AEHD* and *BFGC*; *ABCD* and *EFGH*
2. Three sets: *AB*, *CD*, *GH*, and *EF*; *AD*, *EH*, *FG*, and *BC*; *AE*, *DH*, *CG*, and *BF*
3. *BC*
4. A right angle
5. *AG*, *DF*, *BH*, *CE*

Lesson 41/pages 163-164

1. 684 cm^2
2. 406 in.^{2}
3. 432 in.^{2}
4. 27 ft^2
5. 8,946 in.^{2}
6. 3,060 in.^{2}
7. (1) 598 ft^2
   (2) 1.5 gallons

Life Skill/page 165

(Round to the number of panels needed.)

1. $10\frac{5}{6}$ panels (11)
2. $8\frac{23}{24}$ panels (9)
3. $9\frac{1}{3}$ panels (10)

Lesson 42/pages 166-168

1. 1,080 cm^3
2. 490 in.^{3}
3. 1,728 in.^{3}
4. 27 ft^3
5. 576 in.^{3}
6. 12,000 ft^3
7. 3,300 ft^3
8. 1,093.75 lb.
9. (1) 14.96 gallons
   (2) 124.168 lb.

Lesson 43/pages 169-171

1. The volume of each box is 64 cu cm
2. Box *C*
3. Yes, see problems 1 and 2.

4. 216 in.^{3} or 0.125 ft^3
5. 630 ft^3

Life Skill/pages 172-173

V = 43,200 cu in. or 25 cu ft
The space allotted for food storage is about 21,600 cu in. or $12\frac{1}{2}$ cu ft

Unit 8 Posttest/pages 174-175

1. (2)
2. (5)
3. (3)
4. (2)
5. (3)
6. (2)
7. (4)
8. (2)
9. (3)
10. (5)

Unit 9 Pretest/pages 176-177

1. (1)
2. (4)
3. (1)
4. (3)
5. (1)
6. (4)
7. (3)

Lesson 44/pages 178-179

1. Answers vary.
2. Answers vary.
3. (a) Base
   (b) Base
   (c) Height
4. (a) Base
   (b) Vertex
   (c) Height

Lesson 45/pages 180-183

1. 39.25 in.^{2}
2. 109.90 in.^{2}
3. 149.15 in.^{2}
4. 19.63 in.^{2}
5. 58.33 in.^{2}
6. 77.95 in.^{2}
7. 308.976 cm^2
8. 165.792 cm^2
9. 100.48 or $100\frac{4}{7}$ ft^2
10. 25.3 or $25\frac{19}{56}$ ft^2
11. 30,144 ft^2
12. $204\frac{2}{7}$ or 204.1 ft^2

Life Skill/pages 184-185

5,652 − 565.2 = 5,086.8 in.^{2}

Lesson 46/pages 186-188

(For problems 1-6, use $\pi = \frac{22}{7}$.)

1. 137.5 or $137\frac{1}{2}$ in.^{3}
2. 45.83 or $45\frac{5}{6}$ in.^{3}
3. 138.997 cm^3
4. 416.992 cm^3
5. 100.48 or $100\frac{4}{7}$ ft^3
6. 962.5 or $962\frac{1}{2}$ ft^3
7. 4,227,696 gallons
8. 66.06 gallons
9. 739.2 bushels
10. $314\frac{2}{7}$ or 314.29 ft^3

Life Skill/page 189

1. 502.4 in.^{3}
2. 230.45 in.^{3}

Lesson 47/pages 190-192

1. 3.67 m^3
2. 1,766.25 ft^2
3. 20.5 cu yd

Unit 9 Posttest/pages 193-194

1. (2)
2. (4)
3. (1)
4. (2)
5. (5)
6. (4)
7. (5)

Unit 10 Pretest/pages 195-196

1. (2)
2. (3)
3. (2)
4. (5)
5. (2)
6. (2)
7. (5)
8. (1)

Lesson 48/pages 197-200

1. *F*
2. Any two of these: *ABF, BCF, CDF, DEF, AEF*
3. Pentagon
4. Right pentagonal pyramid
5. Any two of these: *ABFE, BCGF, CDHG, DAEH*
6. Trapezoid
7. Any three of these: *AE, BF, CG, DH*
8. Trapezoidal prism

Lesson 49/pages 201-204

1. Right triangle
2. Either leg, 5 in. or 12 in.
3. 60 in.^{2}
4. 600 in.^{2}
5. 660 in.^{2}
6. 84 cm^2
7. 1,062 m^2
8. 480 in.^{2}
9. 62.4 cm^2
10. 665 m^2

Lesson 50/pages 205-206

1. 36 cm^3
2. 1,890 m^3
3. 604.8 in.^{3}
4. 420 ft^3
5. 60 cm^3
6. 91,394,008.33 ft^3

Lesson 51/pages 207-209

1. 48 packages
2. 331.2 in.^{3}

Life Skill/pages 210-211

1. 9,000 gallons
2. 9 cu ft or 67.5 gallons

Unit 10 Posttest/pages 212-213

1. (3)
2. (2)
3. (2)
4. (2)
5. (1)
6. (4)
7. (5)
8. (5)
9. (1)

Unit 11 Pretest/pages 214-215

1. (2)
2. (1)
3. (3)
4. (2)
5. (3)
6. (2)
7. (5)
8. (1)
9. (4)

Lesson 52/pages 216-217

1. *OA, OC,* or *OD*
2. *CD*
3. 20 cm
4. 6 m
5. 4 cm
6. 28 in.

Life Skill/pages 218-219

1. Between 30° and 45°
2. Between 60° and 75°
3. Between 15° and 30°
4. Between 90° and 60°
5. Europe and Africa
6. 75° and 90°
7. 24,888.9274 miles
8. 24,805.4662 miles

Lesson 53/pages 220-221

1. 1,256 mm^2
2. 314 ft^2
3. 803.84 m^2
4. 5,538.96 ft^2
5. 5,538.96 ft^2
6. 197,279,843 mi^2
7. 30 in.
8. $\frac{1}{4}$
9. $\frac{1}{9}$
10. $\frac{4}{1}$

Lesson 54/pages 222-223

1. 4,186.67 mm^3
2. 523.33 ft^3
3. 2,143.57 m^3
4. 38,772.72 ft^3
5. 38,772.72 ft^3
6. 260,620,153,400 mi^3
7. 30 in.
8. $\frac{1}{8}$
9. $\frac{1}{27}$
10. $\frac{8}{1}$

Lesson 55/pages 224-225

1. 120.11 cm^2
2. 98.13 cm^3

Unit 11 Posttest/pages 226-227

1. (1)
2. (5)
3. (3)
4. (1)
5. (4)
6. (5)
7. (4)
8. (5)
9. (3)